LEADERSHIP ENGINEERING

LEADERSHIP ENGINEERING

Leadership is all about creation

V Srinivasa Prasad

Published by

Invincible Publishers

Published by
Invincible Publishers
201A, SAS Tower, Sector 38, Gurugram – 122003
Phone: +91-124-4034247, +91 9355675555
www.i-publish.in

First Published in 2020

ISBN: 978-93-89600-55-1

Dedicated to my

Mother and Father

for the bliss filled life they blessed me with

ACKNOWLEDGEMENT

Thank you!

I extend my first and profuse thanks to you, my invaluable reader, for picking this from the shelf as a book of your choice. Many contributions helped shape this book to reach it to you.

I, to begin with, must acknowledge with heartfelt gratitude the encouraging feedback, guidance and suggestions provided by Mr B Ashok, CEO Ratnagiri Refineries and Petrochemicals Limited and former Chairman IndianOil Corporation Limited. He has been a great source of inspiration to me.

I am deeply grateful to Prof Soumitra Dutta, Professor of Management and Former Founding Dean SC Johnson College of Business, Cornell University; Chair, Board of Directors, Global Business School Network (GBSN); Founding Editor, Global Innovation Index; Founding Co-Editor, Network Readiness Index; for his acceptance to review my work in my very first and short interaction with him. The analysis and the feedback provided by him were highly valuable and helped me refine the book.

I cannot express enough thanks to Mr DL Pramodh, Executive Director and State Head, IndianOil Corporation Limited, Karnataka for his encouragement and providing the review of the book.

My earnest thanks to Dr Najeeb Kuzhiyil, Oil industry expert at Houston, Texas, USA and author of Spirit of Engineering for his review of the book.

I am immensely thankful to IndianOil Corporation Limited and my colleagues in the company. The culture of creative thinking that IndianOil fostered triggered the interest in me in the subject of leadership.

I must acknowledge with profound thanks the support and contribution I received from the team members of Invincible Publishers in editing, design and formatting and ensuring the successful release of this book.

The stories and anecdotes mentioned in the book are a collection of information gathered from various sources like newspapers, magazines, books, internet, different web pages like Wikipedia, audio books, video documentaries, seminars and conferences etc., The exact source could not be noted down as the initial idea was only a pursuit of interest. However care will be taken to mention the exact sources upon recollections and reminders in the future publications. My deepest gratitude to these and the other anonymous contributors for providing the support and for being the source of information I was interested in.

This book would not have been possible without the support of my wife, son and daughter. I am proud of their love for me and I must, without fail, appreciate their patience.

-V Srinivasa Prasad

PREFACE

Begin with the decision to complete

When I showed the script of this book to my friends, the first question I usually encountered was, 'What made me write a book on leadership?'

This enquiry took my memory back to those years where perhaps, there was no word which disturbed me more than the word leadership. The cognitive agitation swirled by the word leadership continuously kept me unsettled and pushed me to explore the world of leadership deeper. The insights of this exploration emerged out in a form and shape, which is this book. The purpose and the background of this book lie in the vision of 'world of ten million leaders' and also in the fact that leadership is a constant requirement, more so in the current challenging era of machine learning, artificial intelligence, and continuous disruptions.

The book *Leadership Engineering* drives leadership solutions. *Leadership Engineering* is a unique set of concepts and models that let engineer one a leader. Engineering is the application of principles to build or construct something. Leadership Engineering is about the application of leadership principles that have been identified in this book, to build leaders. Individuals, institutions, corporates, and organizations shall be able to engineer leaders for their requirements with the help of this book.

One may wonder if it is really possible to engineer leaders with external inputs. People may think that leadership comes from within naturally as an art. We know artists like painters, actors, dancers, singers, and so on. These artists have art in them. However, not all of them are born artists; many of them are trained artists. Art is an unconscious following of principles. Science is the identification of the principles. In the process of training, the art is converted into science and taught to the learner until it becomes art in him and comes out naturally. Any art is *sciencified* and taught until it is *artified* in the learner. The art of leadership is not any different and *'Leadership Engineering'* does exactly the same; *sciencifying* the art and *artifying* the science of leadership.

Wishing a move forward.

INTRODUCTION

"First thing an idea says 'Hello' to is resistance"

The book *Leadership Engineering* is all about engineering one to become a leader. This book is quite different from all other books on the subject of leadership. The uniqueness of the book lies in its definition of leadership, in its decoding of the leadership process, and in its framework for engineering leaders. The uniqueness of the book also lies in how it brings out facts and concepts realised not so far; about leadership characteristics, differentiation of leadership and other talents, leadership requirements in a corporate, leadership habits, leader societies and in its proposal of a framework for identifying the leadership tasks and leaderees*.

The uniqueness of *Leadership Engineering* is also about being the first to bring out the concept of 'Thought Intelligence' and advocate the strong need for the same and propose tools to imbibe 'Thought Intelligence' in oneself. It attempts to coin new words (*like 'leaderee' as above, the meaning of which and other such coined words are provided in the book at the 'Context') and new definitions to help understand the subject straight to the point.

This book is useful to the corporate and institutions to understand leadership, identify leadership requirements, leadership tasks, and potential leaders, and engineer the leaders effectively to have their leadership requirements met. Leadership Engineering is also useful to individuals in engineering themselves to become leaders. Thus, the target segment for this book includes university students and working professionals for self-leadership engineering, and corporate managers, top brass and HRD practitioners for facilitating leadership engineering in their institutions.

I feel, the word 'leadership' is mostly misused. The first chapter 'Understanding Leadership' argues that it is most important to give a focused definition of leadership to develop the leadership. In the chapter, it provides the most definitive definition of leadership. The definition and the need for the same are explained with the examples of lives of the prominent leaders and personalities.

The characteristics of leadership hitherto not realised are explained in the chapter 'Leadership Characteristics'. In the chapter 'Leadership in Corporate', the leadership needs of the corporate are explained with the unique explanation of the corporate process and life cycle.

In the chapter 'Decoding Leadership Process', the leadership barriers in the process of leadership, the leadership paths, the leadership states and the leadership vehicles required to traverse the path of leadership have been explained with help of narration of the lives of some of the prominent leaders like George Washington and Henry Ford.

In the chapter 'Identification of Leadership tasks and Leaderees', it is dwelled upon defining a leadership task and how to identify the leadership tasks. The techniques to identify the tasks, along with the idea generation techniques and techniques to test the tasks are discussed.

In the chapter 'Engineering a Leader', which is the core of the subject, it is explained how to master the leadership vehicles to cross the leadership barriers like IELS, Diffs, CADFs, Dis-CATs, ReHas to become a leader. The practical techniques to master the leadership vehicles have been presented in the chapter and explained with the help of cases and experiments.

In the chapter 'Leadership Habits', the reader is made to cruise through different leadership habits and the techniques to how to make the leadership habits 'their own'. In the last chapter 'Leader Societies', it is explained how leading societies like the Silicon Valley have a leadership ambiance and have been nurturing leadership, and how to create such leader societies around us.

Leadership Engineering discusses each concept at length. Each idea and concept proposed in the book is the output of an extensive study and analysis of the lives of various leaders and personalities. These concepts are explained with the help of examples of incidents in the lives of the leaders, and with some experiments and cases as well.

Contents

UNDERSTANDING LEADERSHIP

When we say leader or think of a leader, what may come to our mind are thoughts that a leader is a perfect human being, born gifted and exceptional. More often than not, we attach many great things to a 'leader' like perfect, brilliant, extraordinary, highly educated and knowledgeable, smart and so on.

Come on! Breathe easy and relax. This is not what we are discussing in *Leadership Engineering*.

We often attribute qualities to a leader, which are beyond the reach of a common man. In other words, we feel that a leader is not a common man. Yet, on the contrary, history has been telling us that most of the leaders our mankind has seen were common men or women. They were born like any other, grew up like any other or even in conditions worse than most of us. Look at the lives of leaders like Martin Luther King Jr., Mahatma Gandhi, Nelson Mandela, Benjamin Franklin, and Thomas Alva Edison. These undisputed leaders were born and grew up like any common man. So, friends! Let us relax and shun away the idea that leadership is not our kind of job.

If Nelson Mandela had thought so, he would have not created apartheid- free, democratic South Africa. If Mahatma Gandhi had thought so, he would have not created a peace-loving, secular and independent India. If Martin Luther King Jr. had thought so, he would not have created an American society that provides equal rights to the African-Americans.

Sir Li Ka Shing, as you all must be knowing, was born to a teacher in China's Guangdong province. His family took refuge; from southern China to Hong Kong in 1940 due to a war between China and Japan. His father died when Shing was just fourteen. Extreme poverty had forced Shing to leave school and work in a plastics factory. He even had to sell his father's clothes to buy food. He was a common man. However, he went on to create a huge business empire worth more than $154 bn (as of April 2016). He created employment opportunities for thousands of people. He created a booming economy in Hong Kong. He was rated as Asia's richest person worth $31 bn (as of the year 2016) – He created.

The Wright brothers were two of the seven siblings. The brothers never married. In elementary school, Orville was a mischievous boy and was once expelled too. Both brothers attended high school but did not receive diplomas. The family's abrupt move from Richmond to Dayton in 1884 prevented Wilbur from receiving his diploma after finishing four years of high school. The diploma was awarded to Wilbur on April 16, 1994, on his 127th birthday.

In his childhood, Wilbur was struck in the face by a hockey stick while playing an ice-skating game with friends, resulting in the loss of his front teeth. He had been vigorous and athletic until then, and although his injuries did not appear especially severe, he became withdrawn. He had planned to attend Yale. Instead, he spent the next few years largely housebound.

During this time, he cared for his mother who was terminally ill with tuberculosis. Orville dropped out of high school just after his junior year.

Both the brothers had a common-man life. Yet, they were aviation pioneers. They built the world's first successful airplane and its controls which made the aviation industry what it is today. They not only developed the technology but also demonstrated how to use it and built a market and industry around it. They had faced a lot of hardships in the process including life-threatening accidents.

Lee Kuan Yew was born into a wealthy family. He attended law school at Cambridge. He was a leader but not because of this. He was a leader because he founded the independent city-state of Singapore. He built one of the strongest economies in the world. He built from scratch, an industrialised, urbanised, educated, highly safe society which is an inspiration to many nations in the world.

Look at the other personalities like the above, whose work largely benefitted their countries and the world, and whom we hail as leaders like Benjamin Franklin, Bill Gates, Larry Page, Jawaharlal Nehru, Mark Zuckerberg, JRD Tata and so on.

These great leaders have created; they have created something new in which people have seen value. People, societies, sections have been benefitted largely by their creations and hail them as leaders. Their leadership is not about being aristocratic. Their leadership is not about being blue blood or polymath. Their leadership is all about their creations and hence leadership is creation–Leaders create.

Leadership is 'creation'

Leadership is not about being at the helm of the affairs. Leadership is not about being a president of a country. Leadership is not about achieving success. Leadership is not about being talented. Leadership is not being a CEO or MD of a company.

Leadership is all about creation.

We often use the word leader for all high-profile people, political and famous personalities. All management subjects related to human resources are called leadership topics.

There is a need to restrict the use of the words 'leader' and 'leadership' to have a focussed and stressed approach towards the subject. The corporates and institutions often lament that they spend a lot of money on leadership development but still do not get the desired results. This is because the word leadership is used for a very broad purpose and we lack clarity on what to expect from leadership programs. The word leadership is the most misused word.

We give training to our talent in sales skills and call it a leadership development program. We give training to our talent on time management and call it a leadership development program. Mind you, we do not develop leaders by such programs; we develop better sales talent or better resource management talent, for instance. Hence, there is no point in lamenting about the lack of development of sufficient leadership from these programs.

Leadership is not about being an excellent sales talent. Leadership is not about being an excellent boss. Leadership is all different from these. We all have a fair idea about what is leadership and what to expect from a leader. But, we often are unable to express these expectations. We call it leadership if a job is done in an excellent way. We call it leadership if a solution is proposed to a complex management problem. But, all this is quite different from leadership.

Leadership is not about doing something differently or in a better way. Leadership is all about doing something new.

Leadership is all about creating. We all love new things, we all love creation as it increases value. We hail those who do these things for us. We call them great. We call them leaders. Investors want leaders in their companies, expecting that they create wealth for them and add value to their investments. People want leaders in their societies, expecting them to transform their lives. Hence, a leader is wanted and the expectation from the leader is to create; leadership is creation.

Leaders create. They create situations, they create wealth, they create events that had never existed before, they create transformation and hence, leadership is nothing but creation.

As Leadership is creation, is invention a leadership?

An inventor invents new equipment, a technologist designs a new technology, an artist paints a new picture, and a scientist develops a new medicine–are they leaders? Do we ever call them leaders?

We all know Thomas Alva Edison invented the bulb. But, do we know that similar bulbs did exist before? We all remember only Thomas Alva Edison when we think of a bulb. That is because Thomas Alva Edison not only developed a bulb, he created value around his development by showing how the bulb could be put to use. He developed an electricity transmission system. He developed a measuring system for the consumption of electricity and charge for the same.

The automobile was not new before Henry Ford. Aviation was not a new subject before the Wright brothers. The concept of the telephone was not new before Alexander Graham Bell. People before these great leaders even owned patents in these fields. Yet, we remember only these great personalities associated with these fields. We remember them, and they are leaders because they created value around the invention or concept. They brought these inventions to the practical use that benefitted the entire mankind.

The mere invention is not sufficient for the people to make use of it. There shall be a value created around the invention. Talent is required for invention and leadership is required for the creation of useful value of the invention. The relation of creation to the invention is that the creation is putting the invention to mass use.

What is Creation?

To understand creation and to understand why we shall call leadership as creation, we need to understand the features of creation.

Features of creation-

- That which has not existed before
- That which benefits all
- That which lasts
- That which is transformative

Creation is also 'Developing, Making, Founding, Building, Shaping, Forming, Transforming, Rebuilding, Reshaping, Remaking, Revolutionising and Pioneering'

Creation calls for leadership as it is a very important task. Mankind evolved on creation. The creations by leaders have changed the world. The progress we enjoy today is the result of creations. Transportation systems, communication systems, computer systems, and the internet are the creations that have added a lot of value to our lifestyles. The great creations are turning points in the world's history. The leaders who created have never been forgotten. Creation benefits hundreds and thousands of people. Perhaps, creation is the only act that is valued so much. You look at any creation, there is a leader or a leading institution behind that. We need more creations. We need more people who take up this task. We need more leaders.

Creation calls for certain qualities; these are leadership qualities. A person who creates exhibits these qualities. On the path to creation, there are lots of hurdles involved. There are discomforts. There are distractions. There are disappointments. There are fears. There are hardships. The one who overcomes all these and creates is a proclaimed leader. Nobody is bigger than the creation, including the creator, that is, the leader. The fruits of the leadership are not for the leader. Leadership (as leadership is creation) is giving. Leaders excel in giving.

Leadership is the solution for all world problems

Leadership (always read it as creation) ensures progress. It adds and it uplifts. The world today is much better off than that of many generations back. This has been possible due to the leadership acts by the world leaders who have created and transformed. Long back, the world was wrought by superstitions, ignorance, discrimination, racism, oppression, lack of development and technology. The leadership behind the transformation of societies at different points in time and different parts of the world like Europe, America, Africa and Asia with the creation of industries around the telephone, automobile, aircraft, engineering, computer, communications, smart devices, etc. has changed the world for the better and was instrumental for solutions of many world problems. The world is still suffering from four basic problems–poverty, extremism, climate change, and corruption. The leadership with its definition is the definite solution to these basic and associated world problems.

Leadership is different from other talents

We find different varieties of talents around us like artists (musicians, painters, actors, directors and so on), sportsmen, scientists, inventors, technologists, doctors, teachers and the like. Each talent is a profession and hence, can be called professional talent or executive talent. All the talents are very important and find their requirements. When a group of talents (same or different) team up for a common cause, they form a corporate/company/institute. Such corporates require additional talents and these talents are called corporate talents.

- Executive talents (the kinds as listed above)
- Corporate talents
    - Manager
    - Head
    - Management

And then, there is a third talent–the leadership which creates a purpose for all the above talents.

Any talent possesses skills. Talent can be of three intensities–mediocre, good, and great–like average executive, good executive, and great executive; average manager, good manager, and great manager; average head, good head, and great head. But, leadership talent has only one intensity, which is great. We never hear anyone saying that someone is an average leader. That is simply because creation can never be mediocre. The mediocre one can never be a creation.

Management and Leadership

Management and leadership are not (or not necessarily) one and the same. Management requires leaders to fulfill their wishes. Management is the one that seeks results. Leadership is necessarily not a part of management or decision-making.

The management benefits from the leadership. In the corporate hierarchy, management is the stakeholders. Leadership benefits all the stakeholders including investors, customers, employees and vendors and all in the vicinity of the corporate. Similarly, for social leaders management is the public. These leaders work for the benefit of the public.

Heads and Leaders (Headship and Leadership)

People often intermix heads and leaders. It is generally considered that heads are leaders. But, leaders are different from heads.

Before Bill Gates, there was no Microsoft. Conceiving the idea of developing software for microcomputers was an innovation. Bill gates took the idea forward, sold the idea successfully to MITS, established Microsoft, scouted for opportunities, got into agreements with IBM, kept on developing products and thus, built a mammoth company that had never existed before. Microsoft, the great creation of the leader benefitted millions and millions of people, organizations and added tremendous value.

Steve Ballmer took the mantle and carried Microsoft forward. He sustained the company, increased the revenues from $25 bn to $70 bn. Bill Gates created Microsoft. Steve Ballmer continued the growth. It is clearly evident that Bill Gates was a leader and Steve Ballmer was a great head.

Leaders do not have the luxury of the readymade. They make it. They need to see it clearly in their minds and make the world see what they see.

Heads need not make. It is evident before their eyes. Heads take all the pain to increment the value of the creation. Heads take forward the creation of the leaders. Heads ensure that the benefits of the creation continue to flourish at an increased rate. Heads ensure the growth in the benefits of the creation. We have been seeing around us that companies are set up by the founders. They are the leaders in this case. The leaders pass away. Companies still do exist. The leaders create and the heads carry forward. The heads understand the purpose of the creation so well. In the passage of the creation, they ensure that the true spirit is not distorted.

The head is one who succeeds to the highest position in the group. The head becomes *head* due to his experience. People follow him or his instructions due to his position like that of the head of a country, captain of a team and so on. A leader is the one who creates.

Managers and Leaders

We have been observing, for ages, the managers working in the corporates. They keep up the task. The existing one, be it the sales department, be it the production, is maintained so well by the managers to prolong its continuity. Managers nurture and nourish.

I observe in many leadership books, the struggle to differentiate a leader and a manager. It is broadly said that managers plan and organise, etc. and leaders motivate, inspire, etc. I do not understand why there has been always, an attempt to belittle managers in front of the leaders. In fact, even the manager has to inspire, motivate as it is required to be done by a leader. Leadership does more of a convincing job than a motivating job. The details of the leadership-convincing are dealt with in the later chapters. But, managers do need to inspire as they have to get the work done which has been done for long, to sustain the momentum.

Successful people and Leaders

Successful people like actors, musicians, artists, sportspersons, high net worth individuals, heads of the countries rise to the positions with great hard work. Look at the lives of some of the most successful people like Marilyn Monroe, Madonna, Michael Jackson, Pele, Angelina Jolie, Roger Federer; their journeys were never easy. There was a great hard work behind their success. Successful people enjoy a lot of popularity due to their talent, charm, and achievements. They get a lot of recognition in the society, with awards and rewards. They stand as inspirational icons for others to achieve something. Yet, they are only successful or famous or popular personalities.

Walt Disney was the most famous film personality. As a film producer, he has received 22 Academy Awards out of 59 nominations and has won more individual Oscars than anyone else. He was presented with two Golden Globe Special Achievement Awards and one Emmy Award, among other honors. Several of his films are included in the National Film Registry by the Library of Congress.

Walt Disney was a pioneer of the American animation industry; he introduced several developments in the production of cartoons. He set up the Disney Brothers Studio. Walt developed the character 'Mickey Mouse' in 1928, his first and highly successful creation; he also provided the voice for his creation in the early years. As the studio grew, Disney became more adventurous, introducing the synchronized sound, full-color three-strip Technicolor, feature-length cartoons and technical developments in the cameras. The results were seen in features such as *Snow White and the Seven Dwarfs* (1937), *Fantasia*, *Pinocchio* (both 1940), *Dumbo* (1941) and *Bambi* (1942), new animated and live-action films including the critically successful *Cinderella* (1950) and *Mary Poppins* (1964), with the latter one receiving the Academy Award for Best Pictures.

In the 1950s, Disney expanded into the amusement park industry, and in 1955, he opened Disneyland. In 1965, he began development of another theme park, Disney World (now Walt Disney World), the heart of which was to be a new type of city, the 'Experimental Prototype Community of Tomorrow' (EPCOT).

He remains a pioneer of the animation industry and an important figure in the cultural history of the United States, where he is considered a national cultural icon. His film work continues to be shown and adapted, his studio maintains high standards in its production of popular entertainment, and the Disney amusement parks have grown in size and number, attracting visitors from several countries.

Walt Disney created; creation like this makes one stand out as a leader in the crowd of successful people. Mere success is not leadership.

Innovation and Leadership

Innovation is not doing the same thing, again and again, and not doing a thing the same way, again and again. Innovation is doing different things (certainly, better things), or doing things differently (certainly, in a better way). That is why people get tempted to use the catchy phrase 'disruptive innovation'. But, somehow, I prefer using the phrase 'transforming innovation'.

Ideas lead to innovation. Mere having an idea or having innovation skills is not sufficient for the company. The idea must be implemented and put to use to reap benefits. Implementation of an innovative idea is nothing but creation, and it is not an easy task. Only leaders can achieve this; it requires leadership.

Different companies have different forums such as idea conclaves, innovation centres, open house innovation, etc. to generate innovative ideas. Unless there is leadership to put these generated ideas into execution, the ideas will not see the light of the day. The companies shall also have Leadership Engineering programs to generate leadership required to put these ideas into implementation.

Scientists and Leaders:

"In science, credit goes to the man who convinces the world, not the man to whom the idea first occurs."

- Francis Galton

For many centuries, smallpox devastated mankind. In modern times, we do not have to worry about it, thanks to the remarkable work of Edward Jenner and the later developments in his endeavors. Edward Jenner is well-known around the world for his innovative contribution to immunization and the ultimate eradication of smallpox. Jenner's work is widely regarded as the foundation of immunology—even though he was neither the first to suggest that infection with cowpox conferred specific immunity to smallpox nor the first to attempt cowpox inoculation for this purpose.

Scientists formulate theories. Theories help actions. Leaders provide results. Results drive actions. The scientists, who act on their theories that lead to the creation, are leaders.

Gurus and Leaders:

Management gurus and gurus in other fields like scientists; do the research, conduct studies and frame theories that help explain the behaviours and patterns. Companies are built and processes are built on these theories. The theories proposed by the gurus help the leaders build.

Areas of leadership

We find leaders on three fronts-

- Business Leaders
- Social Leaders

(Like political leaders, religious leaders, entertainment leaders, etc.)

- Science Leaders

Scales of Leader

Based on the impact created, leaders can be of different scales as following-

Leaderee:

We often tend to label even a person who just starts the leadership journey or becomes popular in the process, as a leader. All who wish to be leaders are not leaders. They may be potential leaders. They may pass or even fail the tests of leadership. To differentiate a potential leader from a leader, a new word shall be assigned to the potential leaders, which I would like to be called 'Leaderee' if I can take the liberty to coin a new English word.

Leaderee is the one who is identified as a potential leader or is considered for development to a leader. The leaderee may emerge as a leader or may fail. Leaderee becomes a leader if he creates.

Leaderette:

Richard Turere first became responsible for herding and safeguarding his family's cattle when he was just nine. Often, his valuable livestock would be raided by the lions roaming in the Nairobi's National park's sweet savannah grasses, leaving him to count the many losses. At the age of 11, Turere decided it was time to find a way of protecting his family's cows, goats, and sheep from falling prey to the hungry lions.

Turere realized that lions were afraid of venturing near the farm's stockade when someone walked around with a flashlight. He put his young mind to work and a few weeks later came up with an innovative, simple and low-cost system to scare the predators away. He fitted a series of flashing LED bulbs onto poles around the livestock enclosure, facing outward the enclosure. The lights were wired to a box with switches and an old car battery powered by a solar panel. They were designed to flicker on and off intermittently, thus tricking the lions into believing that someone was moving around, carrying a flashlight. And it worked. Ever since Turere rigged up his 'Lion Lights'; his family has not lost any livestock to the wild beasts. Several neighbours of the Turere family in Kitengela have sought Turere's help, asking him to install the system in their enclosures as well. In total, around 75 'Lion Light' systems have been rigged up so far, around Kenya.

The work he put up behind his innovative idea amounts to leadership. He did not invent the LED lights; he did not invent the flickering lights. What he did was the creation of 'lion lights'. The leadership in this was that Turere devised and installed the 'lion lights' and he did all this by himself, without ever receiving any training in electronics or engineering.

Nairobi is the world's only capital with a national park, where wild lions, rhinos and other beasts roam free against the urban backdrop of the skyscrapers rising in the nearby, bustling city center. Each year, thousands of tourists visit the park hoping to catch a glimpse of the lions. But, for the pastoralists and Maasai tribes around the park, the sighting of a lion is usually bad news; valuable livestock is often lost to lions looking for easy prey, prompting rural communities to take matters in their own hands. In some cases, they've killed whole prides of lions that they perceived as a threat, or as retaliation for the lost livestock. The use of pesticides such as Furadan–a tablespoon of which costs less than a dollar and is enough to kill a lion–has become a particularly ruthless way of doing so.

The rising human-wildlife conflict, coupled with a fast-growing urban encroachment, means that Kenya is now home to less than 2,000 lions; a massive drop compared to the 15,000 that lived there just a decade ago. Large sums have been spent in recent years by officials in a bid to protect the lions and strengthen Kenya's tourism industry. Yet, conservationists say that many of these top-down initiatives fail to gain traction due to the local populations. And, this is why creations like that of Turere's homegrown, simple, affordable and effective one can make a big difference.

Creations like this have a huge impact but, are limited to the village or the surroundings of the village or the nearby society or the company or the department within a company. Those who create these small scale creations are micro leaders or leaderettes if I can take the liberty to coin a word, again.

State Leader:

Verghese Kurien, known as the *Father of the White Revolution in India,* whose 'billion-litre idea' of Operation Flood–the world's largest agricultural, dairy development program, made India the world's largest milk producer with about 17 percent of the global output in 2010–11. In the milk-deficient nation, the revolution doubled the milk available per person, within 30 years and made dairy farming India's largest self-sustaining industry with benefits such as increased employment, incomes, credit, nutrition, education, health, gender parity & empowerment.

He helped establish the *Amul* cooperative which is today, India's largest food brand. He founded the National Dairy Development Board (NDDB) to replicate *Amul's* 'Anand pattern' nationwide. He is regarded as one of the greatest proponents of the cooperative movement in the world; his work has lifted millions out of poverty, in India and outside.

Personalities, like Verghese Kurien, whose creations impact the nation are called as state leaders.

Global Leader:

Global Leader is one, whose impact of creation benefits societies, countries, and corporates.

LEADERSHIP CHARACTERISTICS

Leadership is not any godly subject

Leaders are not born. They are not gifted, special or chosen and sent down by god. I recently attended a training program on leadership. Four hours into the course, when the trainer asked us what our understanding of the leader was, one of the participants aptly put it that he had understood by then that anyone could become a leader at any point in time. Although I completely disagreed with the definition of the leadership given in the training program, that participant's remark was highly apt.

Martin Luther King Jr. was a great American leader. He created an American society that was racism-free and provided equal rights to African Americans. King had a lot of humiliating incidents in his childhood. King said that his father had regularly whipped him until he was fifteen and a neighbor had reported hearing King's father telling his son that he would make something of him, even if he had to beat him to death.

King suffered from depression in much of his childhood life. At the age of 12, shortly after his maternal grandmother died, King blamed himself and jumped off a second storey, but he had survived.

In one of the incidents on a bus, he and his teacher were ordered by the driver to stand so that some white passengers could sit down. King refused initially but complied after his teacher informed him that he would be breaking the law if he did not comply with the order. He later verbalized this incident with "the angriest I have ever been in my life". King became romantically involved with the daughter of an immigrant German. King had plans of marrying her but was advised not to do so by his friends who said that an interracial relationship would ignite both, the blacks and the whites. King tearfully told a friend that he could not endure his mother's pain due to the marriage; he broke off the relationship around six months later. He continued to have lingering feelings, as his one friend quoted, "He never recovered."

King had a very common life but went on to create history. He became a leader who greatly impacted American society and mankind.

Leaders can be dumb elsewhere

Jack Ma is a great leader. He created the Alibaba group. Alibaba is a China-based, business-to-business marketplace site that currently serves more than 79 million members from more than 240 countries and territories. Alibaba became one of the most valuable tech companies in the world after raising $25 billion, the largest initial public offering in US financial history. In November 2012, Alibaba's online transaction volume exceeded one trillion Yuan. Jack Ma with the creation of Alibaba, marked a significant step in the development of the internet and e-commerce in China.

Jack Ma had an enormous urge that led to this creation. In his youth, he failed university entrance exam thrice. He was also turned down for jobs, like working at KFC or becoming a police officer. He applied to Harvard for ten times and every time, he got rejected. These never dampened the enthusiasm of this great leader. Because of Ma's success, he was often invited to lecture at famous universities such as the Wharton School of Finance at the University of Pennsylvania, Massachusetts Institute of Technology, Harvard University, and Peking University. In 2005, Ma was selected by the World Economic Forum as a 'Young Global Leader' and was named one of the '25 Most Powerful Businesspeople in Asia' by the Fortune magazine. He was selected by the Businessweek as 'Businessperson of the Year' in 2007 and one of the 30 'World's Best CEOs' by the Barrons in 2008.

Leaders can be dumb elsewhere. Yet, they are leaders because they end up creating. The essential feature of being a leader, whether bright or dumb somewhere; is the task that leads to creation.

What do I want to do/What I want to become?

Leadership stems from the thought of 'what I want to do' rather than that of 'what I want to become'. The leader emphasizes on what he wants to do than what he wants to become.

In the corporate interviews for intake or in the appraisal processes or our chitchat with our colleagues, we always tend to ask what the other wants to become in five years or what their goals are, to become, and so on. One answers happily, without giving much of an effort from the brain, that he wants to become CEO, wants to become MD, wants to become president of the country, etc. If the same person is asked what he wants to do, he pauses for some moments and mumbles an answer.

Parents and uncles ask kids what they want to become when

they grow up. Current-day kids are very smart and ambitious; they say they want to become doctors, engineers, scientists and so on. If the kids were asked the question differently, like what they would do when they grow up, imagine what they would reply.

Nobody can stop anyone to do the work as CEO of a company and contribute, even if one is not. One will be regarded as CEO by the virtue of the work even if one is not a CEO, and not being one does not matter; you will become one if you do the work.

Leadership is not being in the front or on the top

We often tend to call those who are in the front or on the top as leaders. The problem is that we think only these are leaders. Leadership is not being in the front or being at the top. Leaders need not necessarily be at the top position. One is regarded as a leader only by virtue of his creation.

Leadership is more about the task, not so much about the skills

In the quest for leadership, we give more thrust on how to do (the means) than what to do (the task). But, leadership looks for the task first. Leadership tries to get a complete idea of the task, including the strong need for the task. Then, the leadership looks for the means and skills to complete the task. Skills are used only as far as they are useful for the task, but not with an idea to master the skills. Perfecting the skills is never a requirement for leadership.

Leadership has no goals

Leadership has no destination to reach. It is a journey in a creative direction, crossing the milestones. There is no goal in the process of leadership. Leadership does not believe in the goals. Therefore, one should not believe in the goal process since once the goal has been accomplished, it makes the person attain a sense of completion and stop. There is no life after the goal.

Leadership starts with a need. It explores for the task. The leader picks up the task and only the progress is measured against milestones; with a boost-up action plan to make up for the shortfalls. Leadership believes only in progress achieved in the task. Leadership progress has the following process-

Identify the Need- Choose the Task- Decide Milestones- Decide Action plan- Decide Timeline-Measure Progress- Make up for shortfalls with boost up actions- Cross each Milestone- Complete the task- Choose the next task .....

Leadership is not having followers; it is having beneficiaries

"If I were to attribute any single reason to such success that I have achieved, I would say that success would not have been possible without a sustained belief that what I did or attempted to do would serve the needs and interests of our country and our people and that I was trustee of such interests," wrote JRD Tata once, who is the pioneer of Indian aviation, steel, and automobile industries – which have built the image of India internationally, and have economically uplifted millions of Indians who have been commemorating his contributions year after year.

Leaders have people with them and around them. People gather and admire the leaders as they are benefitted; what benefits them is the creation.

Leadership sees no competition

The idea in entire leadership is of doing, not of being. Leadership sees no competition since leadership does not entail a position and leaders do not aspire for a position. A position can be held by only one person. Can we see in any organization, multiple company heads? Can we see in any country, multiple presidents? There is always a contest or selection for head and only one head is selected from a contest.

But, no one can stop any organization to have multiple leaders. There is no contest for leaders. We need more and more leaders. The leadership aspirants do not see a disappointment of not being selected, which the aspirants of headship see. Nothing stops an opposition leader to contribute more than the head of the government in a country. Nothing stops a manager to contribute more than the CEO in a company. As leadership is out of one's own initiative and not out of a selection, it sees no competition.

Leadership is not being better than others

Leadership is not trying to be better than others. It is trying to be perfect. Take, for example, a small activity like reaching to an event on time. Even in spite of reaching late, say by five minutes to the event, you may still be reaching earlier than others as the others could be even more late. You may take comfort in the fact that you are better than others, yet you are late to the event, reaching later than the scheduled time. Likewise is the fact of being the topper in the academics with a score of eighty percent and still being twenty percent short of scoring the maximum. Trying to be better than others certainly shall be a stepping stone but shall not end up being the resting stone. Leaders do not take any comfort in being better than others. They check how far they are from being perfect.

Leadership is not being powerful; it is being useful

Powerful is dreaded, useful is respected. Leadership is being useful. By the definition of leadership, which is creation, leaders are always useful. Many of the powerful personalities in history like Hitler have commanded and controlled the masses but fell short to be leaders. A leader possesses capabilities to be useful not the desire to be powerful.

Leadership is performance–more in mind space:

Performance is an act of doing a task. We keep performing every day, at every stage of life. Performance starts from waking up. How well we wake up, wake up with which frame of mind, and wake up at what time; all these are a part of the waking-up performance.

We perform any task basically in two spaces-

- Real space
- Mind space

Mind space consumes most of the efforts and time. Performance is more of a mind game with exhibiting real skills. Mind game is mostly played within our minds to form opinions and overcome the mental blocks.

Imagine the thoughts that occur while we have to perform the simple task of waking up in the morning. We decide in the mind what time we shall wake up. We try to justify whether and why we should wake up at that time. We doubt if we can wake up at that time because of so many reasons that justify not being able to wake up at that time. We form different opinions about waking up at a particular time and try to overcome various mental blocks to wake up. But the real skill required to wake up is to open eyes, get up and get down the bed.

When we have so much to perform in the mind space for a simple task like waking up in the morning, imagine the complex performance like playing a soccer match and the performance in the mind space involved in it when the performance in the real space is only doing the running, grabbing the ball and kicking it towards the goal.

It is the performance in the mind space that differentiates an achiever from the others and that which is critical in the leadership process.

Leadership is a performance in harmony

The performance takes place in two scenarios-

- Performance in competition
- Performance in harmony

The performance of a player in a soccer match is a performance in competition. The performance of a father at his home is a performance in harmony.

I do not understand fairness in a competition, contest, game, and race. They declare only one as the winner and eliminate all others. I do not understand the objective that such contests achieve. They are counterproductive as they leave many disappointed and only one encouraged. The argument that participation is more important than winning sounds silly and hypocritical. Why declare a winner if participation is more important? Why one should be a winner by defeating another? Why de-motivate many and then subject them to motivational lectures? Competitions are oppressive. The selection process to test the suitability of a person for a particular task is an exception. The selection process is required. Yet, it leaves many discouraged due to the limitation of the number of people it selects for the required task.

Leadership is the solution to all this. Leadership is a performance in harmony. Leadership is the only title won without defeating anybody. Leadership is the solution for the mad rush of competition that the world is experiencing today. You need not take part in any selection process and eliminate others to be a leader. You need not contest elections and defeat fellow contenders to be a leader. The leader looks at high performers as benchmarks to catch up with, but not as competitors to defeat.

Leadership success is completing the performance in harmony

Success is completing the performance; failure is leaving it incomplete. It is not the talent or skill set that differentiates success and failure. It is simply the do-factor. The successful person completes the performance and the failed person leaves it incomplete in midway.

Take the example of Ola and TaxiForSure. It is the same concept that both have; it is the same business model that both have adopted. They both have equal access to talent and technology. Yet, Ola is a success story and TaxiForSure is not so much of that. The difference is that Ola did it and the other did not.

Success is felt in two scenarios-

- Success in competition
- Success in harmony

Success in competition is completing the performance by defeating the other person. Success in harmony is completing the performance by keeping all successful. Leadership success is completing the performance in harmony.

LEADERSHIP IN A CORPORATE

What is Corporate?

Corporate is a set of mutually inclusive beneficiaries coming together to get their needs fulfilled through exchange of the resources they have.

The promoters or entrepreneurs begin with a purpose in mind. The purpose of the promoters is wealth creation. The promoters envisage doing this by extending the service which is the product, to the needy that is the customer. Investors support the promoters, expecting returns. The talent provides the means for the execution and expects income.

- Promoter or entrepreneur: Those who have a purpose
- Purpose: wealth creation, profits, fulfillment
- Investor: One who seeks returns, provides support expecting returns
- Talent: provides the means (Human resources)
- Customer: Beneficiary who has the need
- Product: That which meets the need

The heart of the corporate: Corporate Mutual Need Chain

Corporate Mutual Need Chain is the chain, the links of which are the needs of the corporate members.

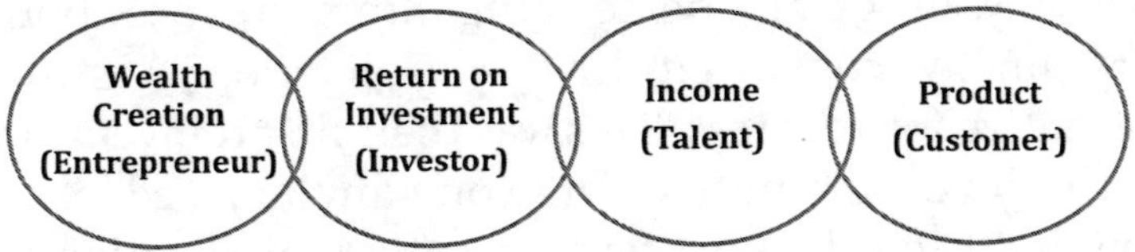

Corporate Mutual Need Chain

The need of the promoter is wealth creation–The need of the investor is the returns – the need of the customer is the product – the need of the talent is the income.

It is needless to mention here that the most important link in the chain is the customer link which is also the most sensitive and the softest link.

Apparent need-Inherent need

Corporate members have two different needs. The apparent need of each member of the corporate seems to be money. But each member has different inherent needs like social status, comforts, security, pride, fame, pleasure, fulfillment, recognition, etc. In the mad rush of chasing the apparent need, people forget their inherent needs, which is the root cause of many problems of mankind. Apparent need alters the priorities.

Corporate Life Cycle

The corporate life cycle consists of different stages in a corporate, which exhibits the attempts and processes to keep the corporate mutual need chain from breaking. When any of the links in the mutual need chain breaks, the corporate ceases to exist or if the need is not getting fulfilled, the corporate dies or reinvents itself.

The telegraph giant Western Union was born in the year 1856 with the merging of New York and Mississippi Valley Printing Telegraph Company and New York & Western Union Telegraph Company. Originally fierce competitors, by 1856, both groups were finally convinced that consolidation was their only alternative for progress. This was their first reinvention to keep up the investor's needs. The company enjoyed phenomenal growth during the next few years. Its capitalization rose from $385,700 in 1858 to $41 million in 1876.

The reinvention in the product link as per the customer needs-

It introduced the first stock ticker in 1866, and a standardized time service in 1870. The next year, in 1871, the company introduced its money transfer service, based on its extensive telegraph network. In 1879, Western Union left the telephone business, having lost a patent lawsuit against Bell Telephone Company. As the telephone replaced the telegraph, money transfer became its primary business.

In 1914, Western Union offered the first charge card for consumers; in 1923, it introduced teletypewriters to join its branches. Singing telegrams followed in 1933, intercity fax in 1935, and commercial, inter-city microwave communications in 1943. In 1958, it began offering Telex service to customers in New York City. Western Union introduced the 'Candygram' in the 1960s.

Western Union became the first American telecommunications corporation to maintain its own fleet of geosynchronous communications satellites, starting in 1974. The fleet of satellites, called Westar, carried communications within the Western Union company; the telegram and mailgram message data to Western Union bureaus nationwide.

Along with satellite telecommunications, Western Union was also active in other forms of telecommunication services-

- The common carrier, terrestrial microwave networks
- Business communications networks such as Telex and TWX, which was later acquired from AT&T Corporation and renamed as Telex II by Western Union
- Landline-based, leased voice and data communication circuits
- Long-distance telephone service
- Airfone air-ground radiotelephone service from 1981 to 1986
- Cellular phone service for a very short time in the early 1980s

Most of these services were discontinued by Western Union in the late 1980s due to a lack of profitability, with the company's the above-said services having been divested and sold to other companies. Again, it reinvented to keep up the investors' needs. Because of deregulation at the time, Western Union began sending money outside the country, re-inventing itself as 'the fastest way to send money worldwide' and expanding its agent locations internationally.

In the later period, Western Union was split into two units– one business consisted of the money transfer business which was funded and operated to take advantage of the significant growth opportunity. The second unit consisted of all the non-strategic communications assets such as the long-distance analog voice network, satellite business, and undersea cable assets. The official name of the corporation was changed to New Valley Corporation in 1991, just in time for that entity to seek bankruptcy protection as part of the strategy to eliminate the overleveraged balance sheet while continuing to grow the money transfer business. Eventually, it fell in the hands of the First Data Corporation. On January 26, 2006, Western Union was spun off as an independent, publicly-traded company with a focus on money transfers. It reinvented itself to keep up the investor needs link and the entrepreneur needs link.

As of February 2006, the Western Union website showed this notice:

"Effective 2006-01-27, Western Union will discontinue all Telegram and Commercial Messaging services. We regret any inconvenience this may cause you, and we thank you for your loyal patronage. If you have any questions or concerns, please contact a customer service representative."

This ended the era of telegrams which had begun in 1851 with the founding of the New York and Mississippi Valley Printing Telegraph Company, and which spanned 155 years of continuous service. It reinvented the product link as per the customer needs by exiting the 155 years of association. Western Union continued to grow in the new business of money transfers with acquisitions and adapting to the latest technological developments like online money transfers and mobile money transfers. The company's revenues stood at USD 5.5 bn in the year 2013.

When the customer need link, investor need link and entrepreneur need link in the corporate mutual need chain were breaking, Western Union reinvented itself to keep all the links together in the chain by divesting itself for the investors' needs and by changing the products for customer needs; disassociating itself from the 155 long years of telegraph association. There are no emotions attached while reinventing to keep up the need chain.

On the contrary, Movie Gallery died within 25 years of its operations. Movie Gallery was formed in 1985 by Joe Malugen and Harrison Parrish in Dothan, Alabama. By June 1987, the company owned five stores and had a franchise operation of 45 stores. In 1988, the company began to consolidate the franchisees into company-owned stores. By 1992, the company had a total of 37 stores and annual revenue of $6 million. In 1999, Movie Gallery announced plans to build 100 new stores. The company completed an 88-store acquisition of Blowout Entertainment in May and ended the year with more than 950 locations in 31 states. In 2000, Movie Gallery again set its goal at opening 100 new stores and relocating 25. This goal was surpassed.

The company moved forward with its largest single-chain acquisition to date, expanding its base of stores by 30% in late December of 2001. The addition of Video Update stores to the Movie Gallery family launched the company's international presence with 100 retail locations in Canada. Following the completion of the Video Update acquisition, Movie Gallery achieved the 2,000 store mark in 2003. In 2005, the company completed the largest acquisition to date with the Hollywood Entertainment merger. This combination of companies increased the store total to 4,700 with revenues of over $2.5 billion. Also, Movie Gallery opened 61 new stores in Western Canada with the acquisition of VHQ Entertainment.

The company began having financial difficulties and announced the closure of 520 stores in September 2007. At that time, Movie Gallery had about 4,500 locations. The next month, the company filed for Chapter 11 bankruptcy protection. Because of these troubles, the stock price dropped below $1 per share and was removed from listing on the NASDAQ stock exchange in November 2007. An additional 400 stores were scheduled to close during the bankruptcy reorganization. Movie Gallery emerged from Chapter 11 in May 2008. In early 2009, Movie Gallery operated about 2,700 locations and 1,300 Hollywood Video locations in the United States. Canadian operations included over 200 'Movie Gallery' branded stores as well as approximately 60 stores under the VHQ brand in western Canada.

The Chapter 11 reorganization plan failed to solve all of the company's problems; Movie Gallery's stock fell from $1.25 in October 2009 to $0.05 per share at the close on December 3, 2009, and many locations fell behind on paying rent. On April 30, 2010, it was announced that all U.S. Hollywood Video stores, Movie Gallery stores, and Game Crazy stores would file for Chapter 7 bankruptcy in May 2010. On June 8, 2010, the company's Canadian stores also entered liquidation. The last US Movie Gallery and Hollywood Video locations closed on July 31, 2010, and the liquidation sale was completed. The remaining Canadian stores were closed during the week of August 8. And, the contents of the company's headquarters were auctioned off in August 2010.

Movie Gallery and its rivals such as Blockbuster have suffered as more consumers switched to services such as Google‘s YouTube, Netflix‘s mail-order service and kiosk operators such as Coinstar‘s Redbox. In the entire life of the company, there was no attempt to reinvent the product portfolio to tune to the changing customers' needs. The company died as the customer-need link broke.

Categories of a corporate

Depending on the stage the corporate is at in the corporate life cycle, the corporate has different stages which are-

- Start-up
- Consolidate
- Keep-up
- Expander

The growth of the corporate is different at different stages.

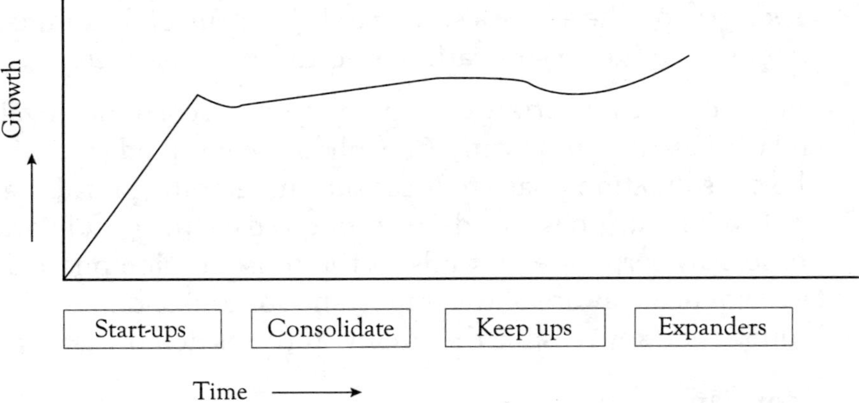

Start-ups:

Corporate starts as a start-up. The activity of creation at this stage is very high. The start-up itself is an activity of creation. Once the corporate comes into existence, it requires creating systems, product, market, finance and accounting system, business model, operations model, etc. In the start-up phase, the corporates grow at a very high rate. Corporates slowly start to feel that they need to consolidate their business as they see the growth is unmanageable and the control is slipping; hence, they see a downfall in the growth.

Many corporates die in the start-up phase itself due to many reasons. One of them is that the mutual need chain breaks quickly or ceases quickly to deliver efficiently.

Consolidates:

Sustainability and stability take the driver's seat in the consolidation phase of the corporates. The companies that undergo this phase can be categorized as consolidates. Consolidates ensure that the actions do not slip out of control.

The consolidates lose some momentum and growth in the process of consolidation without which the corporates face the risk of collapse. In the consolidation stage, corporates seek a grip on the activities, seek to bring control in command systems, and put the operating procedures in place, etc.

Consolidates concentrate on support systems, customer loyalty, and the base to stand firmly. Consolidates aim standardization. The consolidation phase reduces the vulnerability of the start-up. The start-up phase needs to be followed by the consolidation phase. Any corporate that misses the consolidation phase dies fast. Not being aware of the importance of consolidation phase is one of the key reasons that many start-ups do not last long.

Keep-Ups:

The keep-up corporate tends to keep up the growth and their market position that they have achieved so far. Keep-ups look to maintain the momentum and continuity in the presence and in the pace of growth. The keep-ups beam with confidence. At this stage, the corporates tend to improve their efficiency, reduce cost and review the course. Keep-ups continue to do this and look for growth within the same process. Keep-ups give constant results and keep the corporate mutual chain happy for a long, considerable time. The long phase of the keep-up stage brings lethargy into the system and leads to monotony. Meanwhile, the competitors catch up. Keep-ups continue as long as the corporate mutual need chain is intact.

Expanders:

The corporates get into the expanders stage from the keep-up stage when there are strains visible in the corporate mutual need chain. Corporates turn into expanders when the internal measures keeping the corporate mutual need chain from breaking ultimately exhaust. The corporates look to expand their business by entering into new areas of operation, developing new products, and acquiring companies. The growth rate of expanders is higher than the keep-ups. Expanders pass through the hardships more or less like the start-ups.

Leadership needs of a corporate

Corporates need to create continuously. Leaders do exactly the same. Investors prefer leaders since a company with leaders gives better ROI. The leadership needs are different in different phases of the corporate life cycle. The needs of other corporate talents are also similarly different, depending on the cycle the corporate is in.

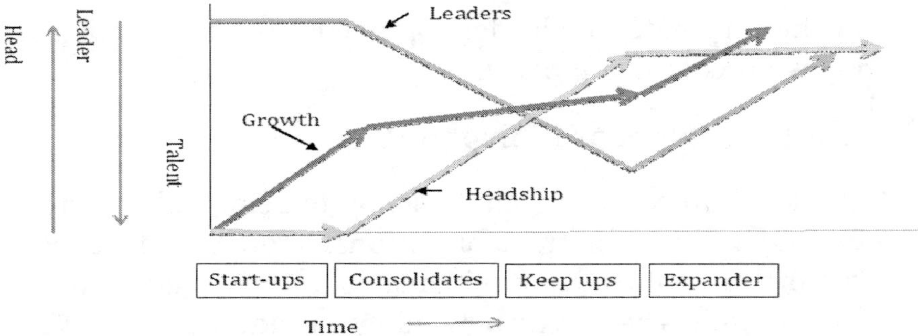

Leadership needs of Start-ups:

The leadership need for start-ups is very high. The founders of start-ups are leaders as they create the start-up. The start-up process requires a lot of leadership tasks like creating products, systems, models, standards, markets, policies, etc. Selling the idea of start-up to an investor is full of hardships. Nobody can see the idea as clearly as the leader. A leader is required to show the people 'the one' which does not exist and seek their support for creating it.

Leadership needs of Consolidates:

The leadership needs of consolidates-corporates decrease as the Consolidates concentrate more on sustenance. The role of headship increases in consolidates. As explained in the earlier section while discussing the difference in leadership talent and headship talent, the headship is most important at this phase to continue the purpose of the start-up, stabilize and make the creation lasting for very long.

Leadership needs of Keep-ups:

Keep-Ups look for incremental growth. This incremental growth meets the growing needs of the corporate. In this phase of the corporate, the leadership needs are very low; headship takes a predominant place.

Leadership needs in Expanders:

Unlike the keep-ups that look for incremental growth, the expanders look for transformational growth and hence, the requirement of leadership increases in the Expander phase as the Expanders have to do new things in addition to the existing, to expand.

DECODING LEADERSHIP PROCESS

Leadership Process is the journey of the leaderee to complete the leadership and the elements involved in it. The Leadership Process contains 'States'. Each State has a path leading to it. The path contains 'Leadership Barriers' which hold the leaderee back. Traversing the path requires 'Leadership Vehicles' to cross the barriers.

George Washington was one of the founding fathers of the United States. He presided over the convention that drafted the current United States Constitution which in turn, devised a new form of the federal government for the United States. During his lifetime, he was called the father of his country. He oversaw the creation of a strong, well-financed national government. Washington's incumbency established many precedents that are still in use today. His retirement from office after two terms established a tradition.

George Washington's interest in the military always showed him the forward path in spite of his early-life struggles. Three of his siblings died before adulthood. His father died of a sudden illness when George was just eleven years old. The death of his father prevented Washington from receiving the education at England's Appleby School, as his older brothers had received. He achieved the equivalent of elementary school education from a variety of tutors and a local school. The talk of securing an appointment in the Royal Navy was dropped for him when he was fifteen years old when his widowed mother objected. Washington contracted smallpox which left his face slightly scarred when he was 19 years old, during a trip with his brother for curing his brother's illness. However, his brother's health failed to improve and he died soon. Washington's interest never allowed these early-life struggles to overwhelm him. His first appointment was of a district adjutant with the rank of Major in the Virginia militia.

George led an aristocratic life in his youth. He was very rich. His richness even increased when he married a rich lady named, Martha Dandridge Custis. He could have been content with life and could have shown indifference to prove more or grow more. But, George had always aspired to prove and grow.

He grew in the ranks starting from obtaining a commission to raise a company of 100 men and starting his military career, to becoming the senior American aide to British General, and to becoming the commander of the Virginia Regiment with the help of his aspiring acts.

In 1758 after the Forbes expedition to capture Fort Duquesne, he retired from his Virginia Regiment commission. Washington did not return to military life until the outbreak of the revolution in 1775.

George Washington's clarity was very high in the times of chaos. His clarity dispelled the doubts and ambiguity in the matters of opposing 1765 Stamp act, 1767 Townshend Acts, and1774 intolerable act. He called out to boycott English goods until these Acts were repealed. He chaired the meeting which called for the convening of a Continental Congress.

After the battles in April 1775, he attended the Second Continental Congress in his military uniform, demonstrating the 'Clarity' that war was imminent. Congress created the Continental Army on June 14, 1775. Washington was then appointed as a full General and Commander-in-chief. He succeeded naturally to the post of General in the American Revolution.

The discomforts, distractions, disappointments, and hardships in subsequent events of his life in the American Revolution were enormous.

The British proclaimed Washington and his continental army as traitors, threatened to confiscate their property and execute them on the scaffold. He lost many of his battles. The appointments in the army by the Congress were a hodgepodge. But, he never surrendered his army during the war, and he continued to fight the British relentlessly till the war's end.

Washington had to face defeat in New York City in August 1776. The Americans were heavily outnumbered, many men deserted, and Washington was badly beaten. Subsequently, Washington was forced to retreat. The future of the Continental Army was in doubt due to expiring enlistments and the string of losses.

Washington's loss at Philadelphia at the Battle of Brandywine in October 1777 prompted some members of Congress to consider removing Washington from command. This movement, however, failed after Washington's supporters rallied in support of him.

The death toll in Washington's army of 11,000 men, in the winters of 1777 at the Valley Forge, numbered in thousands ranging from 2,000 to over 3,000 men (the majority being due to diseases). The winter of 1779–1780 at Morristown represented the worst suffering for the army during the war. The temperatures fell to 16 degrees below the zero, the New York Harbor was frozen over, and snow and ice covered the ground for weeks with the troops again lacking provisions, similar to the time at the Valley Forge. Washington had to face the treason by his associate and a considerable mutiny by Pennsylvania troops in 1780.

All through this, Washington exhibited an enormous 'Passion' to overcome these distractions, disappointments, discomforts, resistance, and hardships; he 'Excelled' to win the war that created a new nation. By the Treaty of Paris in September 1784, Great Britain recognized the independence of the United States. Washington disbanded his army and resigned his commission as the commander-in-chief. The greatest act in his life was his resignation as commander of the army.

After much reluctance, he was persuaded to attend the Constitutional Convention in Philadelphia during the summer of 1787, as a delegate from Virginia, where he was elected in unanimity as the president of the Convention. The new Constitution was subsequently ratified by all thirteen states. The presidency was designed keeping Washington in mind, allowing him to define the office by establishing precedent, once elected.

On being the president, he was guided by an unquenched urge to create a strong, well-financed national government. Washington's incumbency established many precedents that are in use even today. His retirement from office after two terms established a tradition.

The great nation, what it has been, was thus created with the leadership and by overcoming the barriers with the help of leadership vehicles.

The leadership in above and in all other cases follows a typical process called the Leadership Process. The Leadership Process contains basically three elements which are-

- Leadership Barriers
- Leadership Vehicles
- Leadership States

The leader reaches and passes through the 'Leadership States' by facing and crossing the 'Leadership Barriers' that try to hold him back, with the help of 'Leadership Vehicles' that lift him from one State to another. And, this is what the Leadership Process is all about.

The arrangement of the elements in the Leadership Process which are Leadership Barriers, Leadership Vehicles and Leadership States follow a logical sequence called the Leadership Process Line as depicted below. Each element is explained in the subsequent sections-

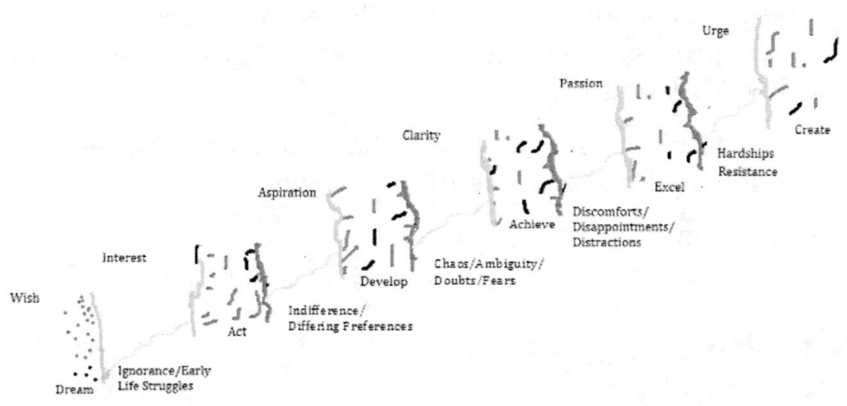

Leadership Process Line

Leadership States

There are different activities that take place in the leadership process. These activities based on their similarities and characteristic nature can be formed into groups. Each group of these activities is distinct from the other group. These groups of activities occur prominently at certain stages or positions in the leadership process. These stages or positions in the leadership process are called 'States'.

For example, in the initial stages of leadership process, the activities of leaderee are mainly related to dreams and imaginations. So, the State that the leaderee is said to be in, at this stage is 'Dream' State. Similarly, at a later point, the leaderee activities will mainly be participation, making moves, acting and doing. This point in the leadership process is called the 'Act' State. The certain point when the leaderee experiences development and the incidents at this stage are better than the mere acts and participation, it is called 'Develop' State. Similarly, there is a stage in the process where the achievement is clearly felt and when the leaderee is seen to be excelling. These States are 'Achieve' and 'Excel' respectively. The point in the process when the leaderee completes the leadership is called 'Creation' State.

A leaderee passes through these different States in the Leadership Process as mentioned below-

- Dream
- Act
- Develop
- Achieve
- Excel
- Creation

The space available at each State is infinite and can accommodate any number of leaderees. The States appear necessarily in the same order as above in the Leadership Process. Each State is connected to the subsequent one by a path. This path contains barriers. It is required to cross these barriers in the path between States, to move from one State to the next.

The leaderee who reaches the last State which is the 'Creation State' is a 'Leader'. Not all leaderees in the Leadership Process pass through all the States to reach the 'Creation State'. Leaderees get stuck up for not being able to cross the barriers. Depending on the State in which the leaderee could reach but is stuck to move further, he is characterized as being-

- Poor (stuck in the Dream State)
- Ordinary (in the Act State)
- Good (in the Develop State)
- Great (in the Achieve State)
- Head/top (in the Excel State)
- Leader (in the Creation State)

Leadership Barriers

I have a childhood friend in my home town. He is a very affluent person. He owns lot of properties around the town. He lives in a huge building which is like a palace. He has many servants, maids and two of them live in the same building to take care. I always wonder the maid in the palatial building and the owner of the building live so near; in the same building and yet, so far. What stops the maid to own such a building like my friend?

We see so many similar examples around. What stops a mechanic in a garage from owning a garage? What stops a chef from owning his own restaurant? Something holds them back. It is the leadership barriers that hold them back. These barriers exist in the path between the States that hold people back at the States and stop their journeys. These leadership barriers are-

- Ignorance (barrier to Act)
- Early Struggles (barrier to Act)
- Indifference (barrier to Develop)
- Differing preferences (barrier to Develop)
- Chaos (barrier to Achieve)
- Ambiguity (barrier to Achieve)
- Doubts (barrier to Achieve)
- Fears (barrier to Achieve)
- Discomforts (barrier to Excel)
- Disappointments (barrier to Excel)
- Distractions (barrier to Excel)
- Hardships (barrier to Leadership)
- Resistance(barrier to Leadership)

We would dwell more on the leadership barriers in the subsequent sections.

Leadership Vehicles

Leaderees require the help of leadership vehicles, embarking which, they cross the barriers for traveling towards leadership. The following are the leadership vehicles that a leaderee requires to travel through the leadership process.

- Wish
- Interest
- Aspiration
- Clarity
- Passion
- Urge

Let us now see how all the elements in the leadership process are interlinked and take the leaderee from one state to the other.

Wish-Dream–Ignorance/Early Life Struggles -Poor

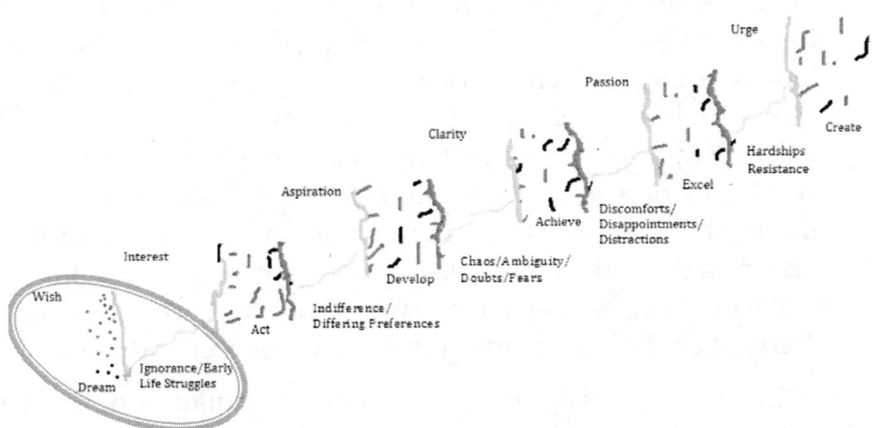

Everyone has a wish. The wish makes leaderee to dream. The journey of leaderee starts with dreams. Wish is the vehicle to 'Dream State'. Nothing happens at this 'Dream State' other than dreaming. The 'Dream State' is thickly populated. A lot of the people get stopped at the 'Dream State'. Leaderees who stop at the 'Dream State' are poor. They all have wishes and dreams. Yet, they do not cross the barriers to reach the next State. The poor feel helpless. They do not know the way forward. The barriers called Ignorance/Early Life Struggles hold back the leaderee from reaching the next State, from the dream State.

Some wish to become rich, some wish to become famous, some wish to become doctors or some wish to become engineers. Different people wish for different things. But, the wish of the leader is the wish to do.

Ignorance/Early Life Struggles (a barrier):

One fine, pleasant month of December, when I was living in Noida, I was going to Nainital; a tourist place in Himalayas, 300 km from Delhi, along with my family. It was a journey of 7 hours by road, specifically by cab. After the initial excitement of the idea of travelling to Nainital, which lasted for two hours, my kids and wife fell asleep. I was watching the road and the driver. He was young and energetic. He clearly looked to be having dreams and wishes. But, he looked not so comfortable. It might be due to the small income that he earned. I knew by then, that there was a barrier which was holding him back from his next State. I started pondering about what could be that particular barrier that would make him remain a cab driver? Likewise, what would be that barrier that would make a bartender remain a bartender and a carpenter remain a carpenter and so on.

My interactions with the cab driver and my analysis of the lives of the similar people who were at this State and the differences in the lives of the people who successfully crossed this phase; have made me know that barriers at this position in the process to the next State are the ignorance and the early life struggles.

Ignorance means 'not knowing' or 'not having the way forward'. This ignorance is different from a lack of knowledge or a lack of awareness. The leaderee is stuck up here, simply because he/she does not know what to do. A person would not become a leader if he is ignorant. He lives in his world and does not know any other world.

Early Life Struggles are of two levels,

a) The initial struggles in the early phase of one's real life,

And

b) The initial struggles in the early phase of leadership life.

The initial struggles in the early phase of one's real-life are like having to start going to work at the early stage in life without going to school in order to bear the burden of the family, losing father/mother and having no caretaker, no schooling, physical or other disabilities, oppression, etc. The initial struggles in the early phase of the leadership life are like having no basic amenities or a conducive environment around.

These barriers have a crushing effect on the leaderee. The leaderee feels helpless.

Interest- Act–Indifference/Differing Preferences-Ordinary

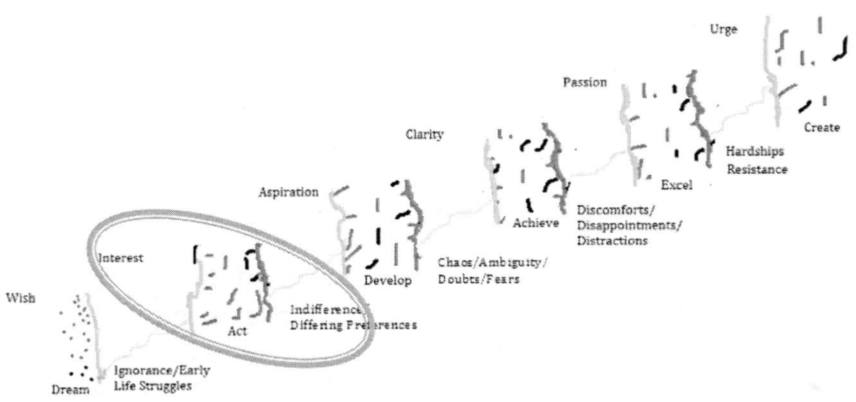

Interest (a vehicle):

Recently, in doing my research, I came across a post on Internet. That was about Erik Carlson, a bartender. He had started as a bartender in San Francisco. He progressed from being bartender to being the Bar Manager and then, the Head of bar operations at one of the iconic bars of Seattle.

As per him, in the interview, he was so miserable and was in the middle of a struggling relationship, and had to take a break before he joined the bar. He was interested in making cocktails and had learned the craft at many bars that he had worked in, in San Francisco. With this art and his interest, he created many drinks which became popular around Seattle. He even named them carefully after some research with the help of Wikipedia.

It was with this interest that this bartender moved to the higher levels in his life. The leaderee requires a carrier called Interest to overcome the Ignorance/Early Life Struggles barrier and move from dream State to work State. Interest as a carrier, makes the leaderee start knowing and thus overcome the ignorance barrier. Interest makes the leaderee pick the task and work for it. This is why we rightly stress on creating interest in the kids in their early childhood or knowing the interest of the kids in their early stage of life.

It is required to trigger Interest in the leaderee at this stage. This subject–How to trigger the Interest–will be dealt with in the next chapter, "Engineering a Leader".

Act (a State):

Interest makes one act. When the leaderee overcomes the ignorance barrier with the help of the leadership vehicle–Interest, he moves to the Act State. Act State is the place where we see the first move, participation, work, and the involvement. Not all move from Dream State to Act State. The Act State is less dense as compared to the dream state, but still crowded. We see many people around who are in the Act State. Due to the next set of barriers, the leaderee finds it difficult to move from this State; the leaderee only moves around, within this State from one point to another point.

Indifference/Differing Preferences

The leaderee encounters barriers called the 'diffs' which are, indifference and differing preferences to move from Act State to the next State.

Indifference (a barrier):

Indifference is to stop showing any concern towards what is required to be done. After the Act State, the leaderee is prone to getting into the comfort zone and may not show any enthusiasm. Indifference comes from satisfaction, lack of ambition and the thoughts such as, "I cannot do more than this", "Why should I do more?" or "I am happy as what I am". Indifference creeps in without our knowledge.

Differing Preferences (a barrier):

My friend who was a Sr. Manager in a software marketing company would always complain about his team members' poor performance in market development. He had three team members who were responsible for sales and market development of the software product. Two of them did exceedingly well in sales, but showed poor results in market development which was pulling down the entire team's performance in this parameter. Market development had a lesser weightage but assumed importance in view of the future prospects. Sales function of this software product was relatively easy as it required being in regular touch with the existing clients, whereas market development required a great deal of efforts and travelling beyond office hours.

One of the two would not do market development as he thought that it would disturb the peace in his life and he wanted a relaxed life. The other would not do it as he thought that the travelling would not allow him to pursue his hobbies. Obviously, they had preferences which were not in alignment with the team's objective. They were unable to contribute to the task.

'Differing Preferences' is doing something other than required. Differing preferences are self-forced or externally forced, for example, you wanted to sit on a weekend evening to read the book Autobiography of a Yogi and your girlfriend or wife wanted you to take her out for a dinner or, the marketing head wants to launch an electric car in the market but his management wants a newer version of the gasoline vehicle to be launched because as per them, the market for electric vehicles is not matured; you would like to work in the area of Artificial Intelligence and Machine learning, however, you have taken up a job in space research as it is a home town posting which would help take care of your ailing parents.

Differing preferences is one of the reasons for the failure of a leaderee in the leadership. The leaderee requires a vehicle called Aspiration to overcome the Indifference barrier to reach the Develop State.

Ordinary (being):

The leaderee who limits his journey to Act State without moving to the next State is called an 'Ordinary'. We see such people commonly, all around. The ordinary people cannot cross the Diffs barriers in the path to the next State. However, the leaderee shall continue his journey and move on from this State.

Aspiration – Develop –Good- Chaos/Ambiguity/Doubt/ Fear

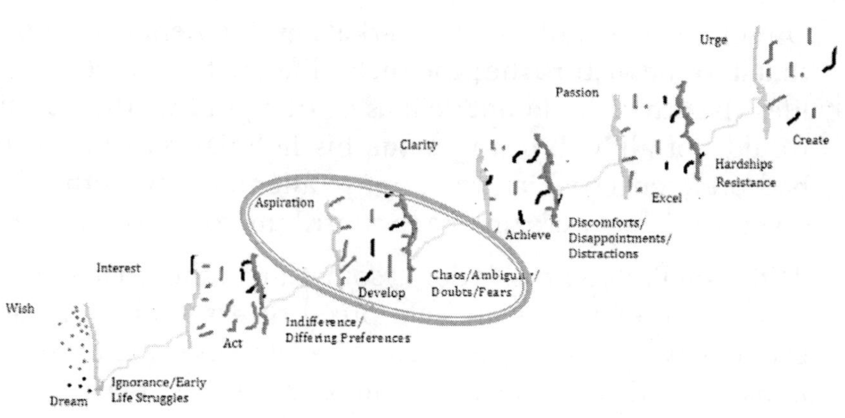

The next State from the Act State is Develop State. The leaderee requires carrier called Aspiration to overcome the Diffs, which are indifference and differing preferences, to move from Act State to Develop State.

Aspiration (a vehicle):

I have this uncle of mine, who is a friend of my dad. He, unlike my dad, would tell that he would have his kids study in the local college, get some degrees, get some job and settle happily and lead a peaceful life. He was thinking about why to take unnecessary pains. "They will be happy if they get any job. Why should I send them to some other place?" He had always thought so. Whereas, my dad had always aspired that three of us, I and my both sisters complete our engineering or medicine and achieve high positions in life. My uncle's two sons settled as clerks in a government school whereas we, the three of us moved on. I completed my engineering and got into an excellent company. My elder sister completed her MBA and has been earning a decent income. My younger sister is a Ph.D. holder and is a principal at a management college. Because of the aspiration my father had, we have seen the development. Aspiration makes the leaderee develop.

It is this lack of aspiration that prevents growth/development. Aspiration provides goals and aims. Aspiration provides direction. Aspiration enables us to orient ourselves to the leadership path and leadership task. Aspiration is a critical leadership vehicle.

Develop (State):

The leaderee with aspiration progresses that is, he reaches the Develop State. Develop State is less dense but you can see many in this State. The leaderee enjoys progress in his deeds. He moves further in his deeds. He betters his position and wealth.

Good (being):

At this stage, the leaderee can be labeled as Good. However, the leaderee is required to continue his journey and move from Develop State to the next State. However, the barriers called CADFs block the leaderee and come in the way of the leaderee.

CADFs (Chaos/Ambiguity/Doubts/Fears) (barriers)

Three management postgraduates, after a grilling selection process, get recruited in a company at the junior management level. Three years on, one of them rises to middle management. The other two are still fighting to come to terms with many things.

We see an ample number of such instances all around. The reasons for such incidents are barriers called Chaos/Ambiguity/Doubts/Fears (CADFs). The path from Develop State to Achieve State contains these barriers.

Chaos:

The leaderee does not see the picture ahead, clearly. He has lot of choices and alternatives. Many paths appear in front of him. He is always confused about which path to take. He is always confused about if the path taken is right. He always wonders if he has a better path. Or the other paths may look better. In the confusion, he leaves the path he is currently on and treads on another path. Suddenly, the priorities get changed. New tasks come up without our control. Sudden opportunities open up in other areas. Unexpected failures take place in the current work area.

Ambiguity:

Ambiguities arise, which are related to purpose, like–what should we do, why should we do, how should we do, whether I should do this or do the other thing?

Doubts:

The doubts that the leaderee has at this stage are related to results and ability such as–Can I, Am I talented enough, will it work, is it good? These doubts linger as no clarification would satisfy the leaderee. We try to seek advice but we still doubt, if these advice are right.

Fear:

Fear is the dreadful imagination of the unknown. In the Leadership Process, the variety of fears that come across are the fear of ridicule, the fear of uncertainty and the fear of failure, like–what if I fail? or if I do this I may fail, so I should not do it.

Clarity-Achieve-Great- Discomforts/Distractions/ Disappointments

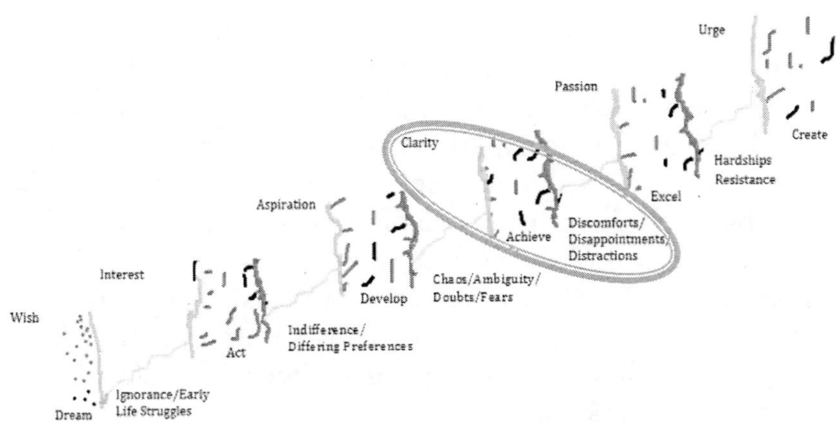

The leaderee requires leadership vehicle called 'Clarity' to overcome the barriers of CADFs to move to the next State.

Clarity (vehicle):

Clarity makes the leaderee overcome these leadership barriers and reach the Achieve State. Clarity provides the leaderee, the direction, and answers many questions. Clarity is the ability to see ahead, very clearly. Clarity enables focus. Clarity enhances confidence. Clarity dispels doubts and fears. Clarity enables the leaderee to not get perturbed by small and short-living issues. Clarity allows focus.

Students prepare for exams as the exams are sure to take place. Tell the students that there might be no exam or, create in them a doubt if the exam will take place at all, and check. No student will be preparing. Even though the idea of preparation is to get a grip on the subject, the students tend to think–let us prepare when we need it.

If you tell the students that the exam is sure to take place and a certain set of questions will come in the exam, all the students including the laziest student will be preparing for the exam and will be doubly sure that he or she has prepared well.

When one is sure what is going to happen, one will certainly act. When one is able to see clearly what is there at the end of the road, one will certainly start walking and reach the end of the road. Let us take an experiment–Tell a group of people to cycle for 20 KM and see how many take up the task and complete it. Then, tell the same group of people to cycle for 20 KM and that at the end of the cycling there is a surprise for them, and then see how many people take up the task. And, if you tell the same group of people to cycle for 20 KM and that at the end of the cycling each one is going to get a free trip to Dubai; see how many people take up the task.

It is the certainty of the outcome of the task that makes people take up the task. If there is any ambiguity or doubt on the outcome, people would not want to take up the task.

Take one more experiment. Prepare four sets of students with ten students per set with a similar age group and a similar pattern of scoring in the exams. Tell the first set of students on an evening, that there is an exam on general knowledge the next day. Check how many prepare for the exam at night before the exam. Tell the second set of students that there is an exam on general knowledge on Europe, the next day. Check how many prepare for the exam at night. Tell the third set of students that there is an exam on general knowledge on the prime ministers of the UK, next day. Check how many prepare for the exam at night. Tell the fourth set of students that there is an exam on general knowledge on the childhood of Winston Churchill, the Great Wartime prime minister of the UK, next day. Check how many prepare for the exam at night.

The experiment reveals that the percentage of students of the fourth set preparing and appearing for the exam is highest of all the sets of students. The clarity that the group has is the highest.

We cannot walk in real space without light. We fumble. We do not know where to go in the darkness. We stop and do not take even a stride or we may move but without knowing where we are going, without knowing what is ahead. We lack direction. Likewise, we cannot move in the mind space without clarity. Obscurity (lack of clarity) in the mind space is like darkness (lack of light) in the real space. In the absence of clarity, we do things without knowing what and why we are doing.

Like the light enables to see (look at) objects in the physical space, clarity provides the ability to see (look at) things in the mind space. The clarity in the mind space is like the light in the real space. Obscurity (lack of clarity) in the mind space is like the darkness in the real space. As we cannot walk in the real space in the darkness, we cannot move in the mind space without clarity.

As the clarity is so important, we shall see how to trigger clarity in a leaderee, in the next chapter.

Achieve (State):

The leaderee, with the help of leadership carrier Clarity, reaches the Achieve State. At this State, the leaderee feels encouraged.

Great (being):

The leaderee who manages to get to this State is 'Great'. 'Greats' have big success to their credit. However, the leaderee may still be forced to be limited to Achieve State by the next set of leadership barriers.

Dis-CATs (barrier):

The next State to Achieve is Excel. Excel is–continuous achievements. In the path to this State, the leaderee is likely to be subjected to distractions, and distress due to disappointments and discomforts. The path to excel contains barriers called Dis-CATs which are Discomforts /Disappointments/Distractions.

Discomforts:

Jeff Bezos was born into an educated family. He graduated with the highest honors from Princeton University with a bachelor of science in engineering degree in electrical engineering and computer science. After graduating from Princeton in 1986, Bezos worked on Wall Street in the computer science field. Then, he worked on building a network for international trade for a company known as Fitel. He next worked at Bankers Trust. Later on, he also worked on Internet-enabled business opportunities at the hedge fund company D. E. Shaw & Co. He was to have a comfortable career and might also have had ended up being a CEO in a very big company.

But, he left his well-paying, comfortable job and trod a path that was having all discomforts. He started a company in his garage. He started his company as an online bookstore and later diversified it to sell DVDs, Blu-rays, CDs, video downloads/streaming, MP3 downloads/streaming, audiobook downloads/streaming, software, video games, electronics, apparel, furniture, food, toys, and jewelry. Thus, he founded Amazon.com which is now the largest online shopping retailer in the world and one of the first of its kind.

The leaderee in his journey to leadership comes to face a lot of discomfort. These discomforts can be like leaving physical luxuries, entertainment, the luxury of leisure time and idling. The leaderee is required to sacrifice his personal life.

Distractions:

Leaderee comes across many distractions. Distractions are temptations in the other fields. These distractions lead to taking up some other task, leaving the current task behind. Taking up another task is like starting afresh which takes its time and effort. The distractions make the leaderee land in some another plane, altogether. Distractions are caused by disappointments. Distractions lead to digressions and do not allow the leaderee to excel and reach the top.

Some of the distractions are like your colleague doing something new and getting a sudden success and you get tempted to do the same leaving your current task, or you are interested to work in the field of Robotics and you get an offer to work in the space research with very high salary. The distractions could be new opportunities in other fields or the temptations arising from the success of peers in other fields, for example, you have started your own company and after two years, you are waiting for something remarkable and your friend with similar capabilities to you has grown in these two years in the company where he has taken up a job and is slated to be the next CEO in his company, etc.

Distractions consume time and hamper progress. The effect of distractions on progress and the time taken in the leadership task is illustrated below in the curve called Progress-Time-Distraction curve.

Progress-Time-Distraction curve

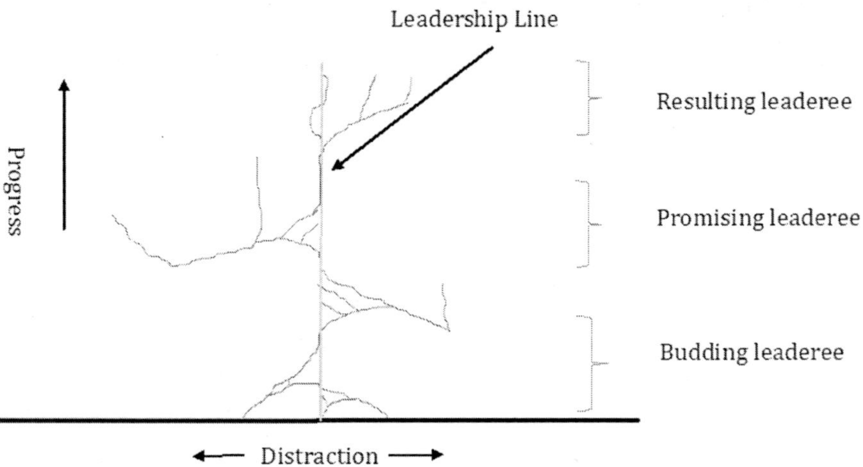

In the Progress–Time-Distraction curve, Y-axis is progress. X-axis on either side of the leadership line is distraction. The length of the progress trajectory is time. The angle of the progress trajectory with Y-axis is the effect of distraction.

The effect of distraction is nil if the angle is zero and the effect of distraction is maximum if the angle is 90°. If the angle is more than 90°, the effect of distraction is negative progress. The measure on the X-axis of the progress trajectory is the net distraction. The measure on the Y-axis of the progress trajectory is the net progress. The length of the trajectory is the actual time taken to reach the point in the curve.

Distractions let leaderee deviate from the leadership line and take another line. The leaderee realizes the distraction and comes back to the leadership line. In the process, he loses time. Some leaderees do not realize the distraction and can never come back. The shortest time taken to complete the leadership line is when there is no distraction.

Disappointments:

Disappointments have two effects–Stopping and pulling down. The disappointments are caused by a lack of support to your views, no short term success, setbacks, hurdles, etc.

Pullela Gopichand has been creating a sensation in Indian Badminton. He has been producing champions who are constantly bringing trophies and medals to India. Before Gopichand, Badminton in India–which had been a Cricket dominating nation–was not given any importance. There were very few players who won any International championships. There were no considerable facilities for training. Gopichand faced many initial struggles to have proper training. He went on with Badminton and won many championships with All England Open as his best performance. He wanted to create a world-class training facility to remove the initial hurdles and provided the opportunities to the younger aspiring talent.

He did not have sufficient funds. His academy, built on the land provided by the Government and with the funds provided by enthusiasts, was just half-built. He made numerous presentations to corporates for funding. People, who otherwise looked at him as a role model, took his autographs and pictures with him, started avoiding him and stopped attending his calls. He was faced with numerous incidents of disappointments, like having to wait outside the chairman's office for three days only to be told that Badminton had no charm to draw funds. Frustration was mounting.

Dis-CATs (discomforts, disappointments and distractions) pull down the leaderee and crush him to the ground. Leaderee requires the leadership vehicle called 'Passion' to overcome the Dis-CATs.

Passion-Excel-Head/Top–Resistance/Hardships

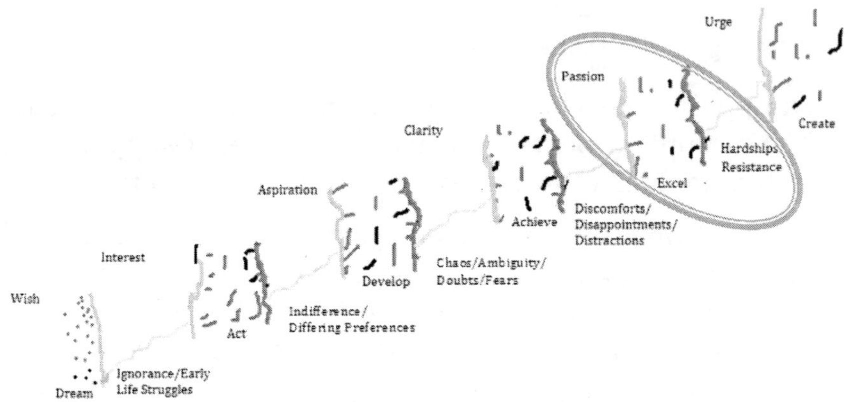

Passion (vehicle):

Continuing from the previous section, 'passion' drove Pullela Gopichand to overcome the Dis-CATs. It was this passion towards his set task that made him even ready to sell his own house to fund building his academy. It is passion that made him work from 4.00 AM to 7.00 PM in a day. It is passion that had made him keep his family comforts aside for the task. It is passion that had made him withstand the criticism from media and other players. It is his passion that had made him overcome the disappointments, like some medal-winning students leaving him to go to other coaches. It is his passion that did not allow him to get distracted by retirement thoughts after the first few medals won by the students of his academy.

The leaderee is required to possess passion to overcome the Dis-CAT leadership barriers. Passion makes the leaderee relentless, come what may. Passion is the fire that requires no fuel once triggered. Passion is having an angle of zero degree with the Y-axis in the Progress–Time-Distraction curve shown above. No distraction disturbs a passionate person.

Excel (State):

At this State, the leaderee continues to achieve much and much. These achievements are frequent and never-ending. The excel State is occupied by very few people. Like any other State, the space available in this State is unlimited. The opportunities abound for anyone, to excel.

Head/Top (being):

One who reaches the excel State and limits himself to this State is Head/Top. He reaches the highest position (top position) in his profession. We find CEOs, heads of the companies, heads of the groups, etc. at this State. The popular reckoning is that one who reaches this position is a leader. However, after going through the definitions provided thus far in Leadership Engineering you now appreciate that leadership is not about being a head. Leadership is something else. The popular misbelief that headship is the leadership is the very reason that most of the heads/tops do not venture to move to the next State which is creation (leadership) State.

Resistance/Hardships

The next State to Excel State is Creation State. The path for one to convert himself from excel State to creation State contains barriers called Resistance and Hardships.

Resistance (barrier):

The first thing an idea says hello to is resistance. Resistance is a continuous and repeated occurrence in leadership. The creation is all new. Nobody has seen it before. There is always a lot of resistance to something which is new. Most of the people around turn to be disbelievers. Resistance also emerges due to conflict of interest, perceived by a smaller group of people, against larger benefits.

Hardships (barrier):

Nelson Mandela was one of the greatest leaders that the 20th century had witnessed. He created apartheid-free South Africa. His path was full of hardships. He spent 27 years of his life in jail. He was banned many times from public appearances and was arrested many times. These hardships never deterred him as his urge for apartheid-free democratic South Africa was insurmountable.

The path to creation is full of hardships. The disbelievers and the resistance to anything new cause lot of hardships for the leaderee to create.

Urge-Creation-Leader

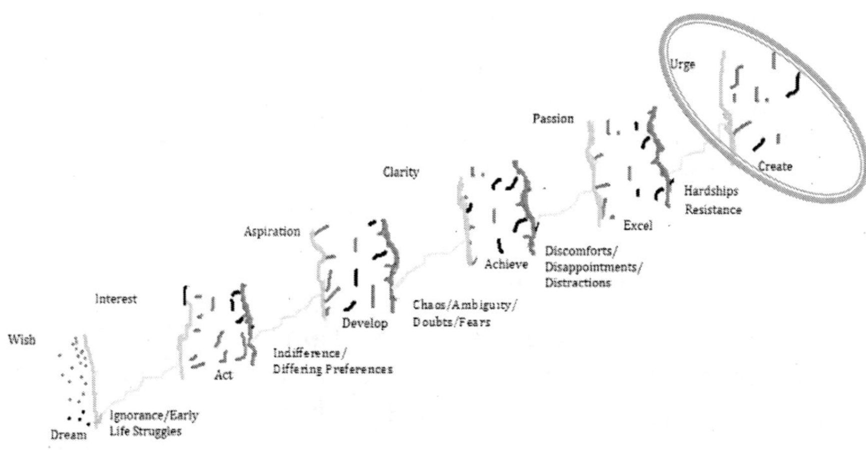

Leadership Urge (vehicle):

Abraham Lincoln, the 16th president of USA, transformed America into a slavery-free nation and thus, changed the outlook of the entire world. He led America through its bloodiest civil war, preserved the union and modernized the economy. His fight against slavery had worldwide impact and led the world view that all men are born equal.

Abraham Lincoln exhibited enormous urge for the cause. It was the urge in this largely self-educated, not so popular person which got him from nowhere to be the President of USA and enabled him to transform the nation.

Whereas Stephen A Douglas, who could have made the world not know Abraham Lincoln, was the opponent to Lincoln in the Presidential run in the year 1860, he had already defeated Lincoln in the race for US Senator from Illinois and was even a stronger and more popular presidential candidate than Lincoln, was highly educated, had powerful friends, had a promising career in politics as a President. However, he had only the desire to be the President; he lacked the urge that Lincoln had. Douglas could not take a position in the critical times and could not swing the people as Lincoln did.

Immediately on Lincoln's election, the separatists from the southern states supporting slavery decided to leave the union and the threat of US splitting loomed large. The worst civil war in the history of US broke out.

Lincoln's urge on his emancipation of slavery and keeping the USA united made him stand firm throughout the crisis, face the insurmountable resistance and hardships, get support even for the second term and emerge victorious in creating the transformed USA.

The urge is a deep desire. Leadership urge is a deep desire for creation. Urge keeps the leader going and overcome the resistance and hardships. Leader with urge sees no pain in the hardships. Leaderee requires the urge to turn into leader. We shall see how to trigger urge in a leaderee, in the next section.

Creation (State):

Leaderee who embarks the leadership vehicle urge, reaches the

'Creation State'. The leaderee completes his task and ends up creating. The Creation State is the ultimate State in the journey of leadership. The Creation State is sparse in the current world and there is lot of room for many in this State.

Leader (being):

In the Creation State, the leaderee completes his leadership journey and turns into leader.

The story of Henry Ford:

Henry Ford was one of the outstanding leaders mankind has ever seen. He revolutionised automobile industry. He created this great company called Ford Motor Company. He developed and manufactured the first automobile that middle-class Americans could afford. He developed an assembly line technique for mass production which had a profound impact on the landscape of the twentieth century. However, his journey was never smooth. He had all the barriers. He overcame all of them in the journey of his mammoth creation.

Early Life Struggles and the Interest:

Ford was devastated when his mother died in 1876; Ford was just thirteen years old. His father expected him to, eventually, take over the family farm. Ford attended a one-room school for eight years when he was not helping his father with the harvest. But he despised farm work. He later wrote, "I never had any particular love for the farm—it was the mother on the farm I loved."

At age 16, he walked to Detroit to find work in its machine shops where he came in contact with the internal combustion engine for the first time. In 1882, he returned to Dearborn to work on the family farm. Ford had to support his family, wife and one child by farming and running a sawmill, where he became adept at operating the Westinghouse portable steam engine. He was later hired by Westinghouse to service their steam engines.

Ford developed an interest and pursued it wherever opportunity lay–starting from dismantling and reassembling the pocket watch his father had gifted him, being adept in operating steam engine on the farm, eventually he built a small 'farm locomotive', a tractor that used an old mowing machine for its chassis and a homemade steam engine for power.

The clarity in the chaos:

Throughout the chaos of his life, he never had an ambiguity about what to do. His clarity on the idea of developing an automobile was great and evident in the pursuits of his experiments on gasoline engines in the spare time at Edison illuminating company.

Achievement:

Ford got his first success in Edison Illuminating Company. After his promotion to Chief Engineer in 1893, he had enough time and money to devote attention to his personal experiments on gasoline engines. These experiments culminated in 1896 with the completion of a self-propelled vehicle which he named the Ford Quadricycle. After various test drives, Ford brainstormed ways to improve the Quadricycle where he was introduced to Thomas Edison. Thomas Edison, after being introduced to Ford, approved Ford's automobile experimentation. Encouraged by Edison, Ford designed and built a second vehicle, completing it in 1898.

Discomforts:

Ford Left the comfort of a successful career at Edison Illuminating Company and ventured to tread on an unknown path, of founding a company.

Disappointments:

He left Edison and founded the Detroit Automobile Company on August 5, 1899. However, the automobiles produced were of lower quality and of a higher price than what Ford wanted. Ultimately, the company was not successful and was dissolved in January 1901.

Ford formed the Henry Ford Company on November 30, 1901 along with Murphy and other stockholders with himself as the chief engineer. In 1902, Murphy brought in Henry M. Leland as a consultant; Ford, in response, left the very company that bore his name.

Ford received the backing of an old acquaintance, Alexander Y. Malcomson, a Detroit-area coal dealer. They formed a partnership, 'Ford & Malcomson, Ltd.' to manufacture automobiles. Ford went on to design an inexpensive automobile, and the duo leased a factory and contracted with a machine shop owned by John and Horace E. Dodge to supply over $160,000 in parts. Sales were slow, and a crisis arose when the Dodge brothers demanded payment for their first shipment.

Distractions:

The successful career in Edison illuminating company which lifted Ford's financial status could have been a big distraction to Ford. But he never got distracted.

Excel:

Ford also produced the 80+ horsepower racer '999' which Barney Oldfield was to drive to victory in a race in October 1902. Ford then demonstrated a newly designed car on the ice of Lake St. Clair, driving 1 mile (1.6 km) in 39.4 seconds and setting a new land speed record at 91.3 miles per hour (146.9 kilometres per hour). Convinced by this success, the race driver Barney

Oldfield, who named this new Ford model '999' in honour of the fastest locomotive of the day, took the car around the country, making the Ford brand known throughout the United States.

Hardships:

The Ford Motor Company was incorporated on June 16, 1903 with $28,000 capital. The company was a success from the beginning, but just five weeks after its incorporation, the Association of Licensed Automobile Manufacturers threatened to put it out of business because Ford was not a licensed manufacturer. He had been denied a license by this group, which aimed at preserving for its members, the profits of what was fast becoming a major industry. The basis of their power was control of a patent granted in 1895 to George Baldwin Selden, a patent lawyer of Rochester, New York. The association claimed that the patent applied to all gasoline-powered automobiles. Along with many rural Midwesterners of his generation, Ford hated industrial combinations and Eastern financial power. Moreover, Ford thought of the Selden patent as preposterous. All invention was a matter of evolution, he said; yet Selden claimed genesis. He was glad to fight, even though the fight pitted the puny Ford Motor Company against an industry worth millions of dollars. The gathering of evidence and actual court hearings took six years. Ford lost the original case in 1909; he appealed and won in 1911. His victory had wide implications for the industry, and the fight made Ford a popular hero.

During its first five years, the Ford Motor Company produced eight different models, and by 1908, its output was 100 cars a day. The stockholders were ecstatic; Ford was dissatisfied and looked towards turning out 1,000 a day. The stockholders seriously considered court action to stop him from using profits to expand. In 1909, Ford, who owned 58 percent of the stock, announced that he was only going to make one car in the future, the Model T. The only thing the minority stockholders could do to protect their dividends from his all-consuming imagination was to take him to court, which Horace and John Dodge did in 1916.

The Dodge brothers, who formerly had supplied chassis to Ford but were now manufacturing their own car while still holding Ford stock, sued Ford for what they claimed was his reckless expansion and for reducing prices of the company's product, thereby diverting money from stockholders' dividends. The court hearings gave Ford a chance to expound his ideas about business. In December 1917, the court ruled in favour of the Dodges; Ford, as in the Selden case, appealed, but this time he lost. In 1919, the court said that, while Ford's sentiments about his employees and customers were nice, a business is for the profit of its stockholders.

Create:

"I will build a motor car for the great multitude," Ford proclaimed in announcing the birth of the Model T in October 1908. In the 19 years of the Model T's existence, he sold 15,500,000 of the cars in the United States, almost 1,000,000 more in Canada, and 250,000 in Great Britain, a production total amounting to half the auto output of the world. The motor age arrived owing mostly to Ford's vision of the car as the ordinary man's utility rather than the rich man's luxury. Once only the rich had travelled freely around the country; now millions could go wherever they pleased. The Model T was the chief instrument in one of the greatest and most rapid changes in the lives of the common people in history, and it effected this change in less than two decades. Farmers were no longer isolated on remote farms. The horse disappeared so rapidly; the transfer of acreage from hay to other crops caused an agricultural revolution. The automobile became the main prop of the American economy and a stimulant to urbanization—cities spread outward, creating suburbs and housing developments—and to the building of the finest highway system in the world.

The remarkable manufacturing rate of Model Ts was made possible by the most advanced production technology yet conceived. After much experimentation by Ford and his engineers, the system that had evolved by 1913–14 in Ford's new plant in Highland Park, Michigan, was able to deliver parts, subassemblies, and assemblies (themselves built on subsidiary assembly lines) with precise timing to a constantly moving main assembly line, where a complete chassis was turned out every 93 minutes, an enormous improvement from the 728 minutes which were formerly required. The minute subdivision of labour and the coordination of a multitude of operations produced huge gains in productivity.

In 1914, the Ford Motor Company announced that it would henceforth pay eligible workers a minimum wage of $5 a day (compared to an average of $2.34 for the industry) and would reduce the work day from nine hours to eight, thereby converting the factory to a three-shift day. Overnight Ford became a worldwide celebrity. People either praised him as a great humanitarian or excoriated him as a mad socialist. Ford said humanitarianism had nothing to do with it. Previously, profit had been based on paying wages as low as workers would take and pricing cars as high as the traffic would bear. Ford, on the other hand, stressed low pricing (the Model T had a cost of $950 in 1908 and $290 in 1927) in order to capture the widest possible market and then met the price by volume and efficiency.

Ford's success in making the automobile a basic necessity turned out to be but a prelude to a more widespread revolution. The development of mass-production techniques, which enabled the company eventually to turn out a Model T every 24 seconds; the frequent reductions in the price of the car made possible by economies of scale; and the payment of a living wage that raised workers above subsistence and made them potential customers for, among other things, automobiles—these creations changed the very structure of society.

Take any leader; who has created has essentially undergone similar process.

IDENTIFICATION OF LEADERSHIP TASKS AND LEADEREES

"Convincing is showing"

Leadership tasks in a corporate

Any task is leadership task–as long as it ends up in creation. The performer of such task is a leader. For any company to grow regularly and tackle their ups and lows, the company should be aware of the leadership tasks that it needs to embark upon.

The tasks in any corporate work are basically, of three varieties-

- Executive tasks
- Headship tasks
- Leadership tasks

Executive tasks are such as completing a sales transaction, completing purchase process or a thread in the process, payment process, training process, recruitment process, providing

IS support, arranging logistics, and completing one's part in the manufacturing. These tasks are must in any company for running the current day's show.

Headship tasks are such as resource planning, targets, forecast, controls and checks, developments, directions, strategies, facilitation, coordination etc., which are must in a company for the direction and progress of the company. It goes without saying that executive tasks contain headship tasks limited to the area of operation, and headship tasks contain executive tasks too, in a limited way. It is the dominance of the nature of the tasks that decides whether it is an executive function or a headship function.

Leadership tasks

One may question that–As leadership is defined as creation, once a company is created, is there any scope for further creation within the created? The answer is yes. There is always a huge scope of the creation within a company that has already been created.

Tata group, one of the most successful groups of companies and biggest Indian multinationals, was founded by Jamsetji Nusserwanji Tata in the year 1868. He created path for industrialization in India by seeding pioneering businesses in sectors such as steel, energy, textile, hospitality, etc. He created a group and brand that was to last for very long, that was to be pride of India and that was to inspire India, transforming it into an industrial nation.

The next chairman of the Tata group, Sir Dorabji Tata built first-ever steel plant in India and created Tata Iron and Steel. He broke a new ground by building a hydroelectric power plant. He built an institute, Indian Institute of Science, a premier institute that would create brightest talents in India. He created a trend and inspired the company–that would build many such institutions later, like Tata Institute of Fundamental Research,

Tata memorial centre, JRD Tata eco-technology centre, etc. Sir Dorabji created a culture in the company that would take good care of their large family of employees. He created 8-hr working day system, townships that were not heard of before and emulated by other corporates. Sir Dorabji created new business areas for the company like insurance, soaps, and detergents, cooking oil, etc.

JRD Tata became chairman of the Tata group in the year 1938. During the period of more than five decades of being at the helm, he created new spheres of business. He created Tata aviation service which would subsequently become nationalized and become India's national carrier–Air India. He created new companies like Tata Chemicals, Tata Motors, Voltas, Tata consultancy services, Tata tea, and Titan industries.

Ratan Tata took over as the chairman after the retirement of JRD Tata. He created a 'Tata' that is global and multinational. He created a new image of the group by going global through acquisitions of Tetley, NatSteel, a controlling stake in the Brunner Mond group, Corus and Jaguar Land Rover. He created newer business avenues in telecom by setting up Tata Teleservices and acquiring VSNL subsequently. By conceiving and successfully producing Tata Nano, he created an ambience in India where even low-income families started dreaming and soon realising the dream of owning a car.

Tata is a company which created within creation with the help of continuous leadership and proved that leadership is indeed, a continuous requirement in any corporate.

The following leadership tasks in a corporate make you readily appreciate the need for continuous leadership-

- Creation of new business process
- Creation of new business avenues
- Creation of new verticals

Typical Leadership tasks that exist in a corporate are–Create, Develop, Make, Found, Build, Shape, Form, Transform, Rebuild, Reshape, Remake etc. Some of the leadership tasks that can exist in different departments of a company are presented below to have the ready understanding and acknowledgement.

Purchase department:

Purchasing without Kickbacks – Creating a system for that

Creating a procurement system

Creating a vendor base

Creating a Negotiation and base price system

Creating a budgeting system

Creating a price movement tracking system

HR department:

Creating a new recruitment system

Creating a *Leadership Engineering* model in HRD

Devising innovative trainings

Creating systems for most effective utilisation of human resources

Sales department:

Channel creation

Sales Model creation

New Market Creation

Marketing department:

Building brands

Creating new markets

Creation of marketing model

Business Development department:

Post-merger Integration

It is left as food for thought for the reader's imagination to identify the leadership tasks that are possible in the departments like Finance, Production, IS, Management systems, etc.

Framework for Identification of Leadership tasks

Those who do not have a task in mind can accomplish nothing. Even those who have a task in mind may not necessarily accomplish.

Identification of the task is the first step in the Leadership. Individuals, corporate, societies, countries should have a task clearly cut out in mind. However, having the task in mind does not ensure the completion of the task. It requires leadership. Individuals, corporates, societies may have pre- identified tasks or allow the leaderees to identify the task that can benefit them.

There is a strong need for a framework for identification of the leadership tasks for an individual, for a corporate or for a society. In the absence of such framework, the leaderees may get misguided in the enthusiasm of leadership and may end up believing the task they envisage is a leadership task. They may end up pursuing wrong tasks that may lead to waste of resources of the company and eventually ruining the careers of the leaderees. Framework for identification of leadership tasks ensures that no crazy idea is pursued by any leaderee to spoil his/her career. Framework for identification of leadership tasks ensures proper checks in determining the leadership tasks.

Framework for identification of leadership tasks has the following blocks

N–E–T–E–T

- Identify the ***N***eeds or Improvements
- Examine the ***E***xisting Setup (if needs/improvements can be met with the same)
- Conceive the ***T***ask
- Establish the ***E***ffectiveness of the task
- ***T***est whether the task meets the need

The logical flow of the blocks can be pictorially represented as below.

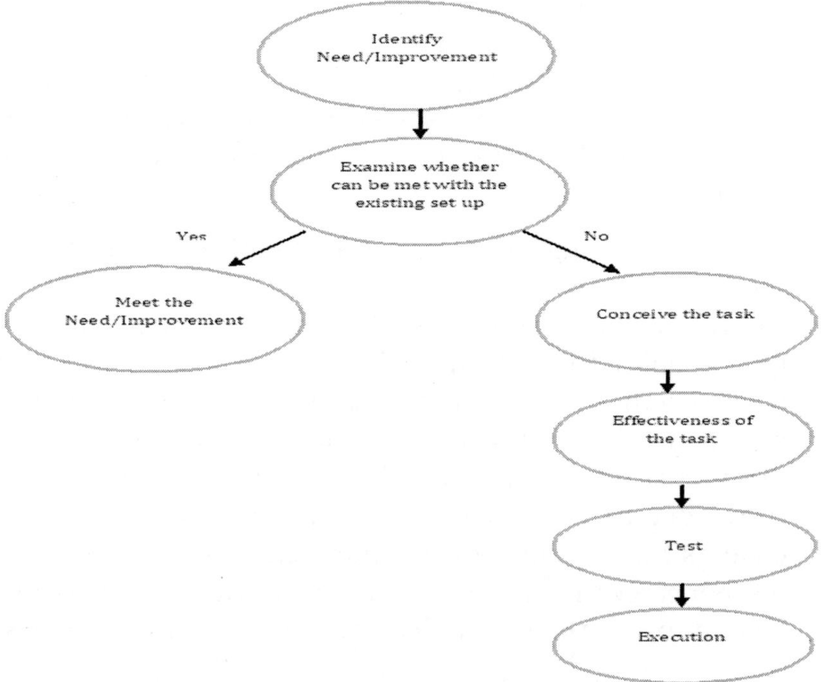

Identify the needs and improvements

Need if not met threatens the survival. Improvement betters

the existence. Wearing woollen clothes in the winter is a need and wearing right colour of the woollen clothes that enhances one's looks is improvement, to cite an example.

Visualising a need is easier than visualising improvement. It is precisely for this very reason or limitation that most of the corporates or individuals do not act until it becomes a need. The corporates or individuals, who act before it becomes a need, do prosper; those who act when it becomes a need just survive; and those who never act perish.

A need forces itself to be identified most of the times. Take for example, the needs like controlling the climate change, reining the terrorism, controlling the corruption and alleviation of poverty which are the four major and most urgent needs of the human race; needs like these are most evident and clear. However, sometimes we may not be able to notice certain needs and requirement of improvements. We need mechanisms to know and understand such needs and requirement for improvements.

Development Cells:

Development cells in the organisational setup of the corporates can be of use in identifying the needs and improvements that the corporate can pursue. Such development cells should be exclusive and should have the constant quest to identify the needs and improvements.

8hr16mnts:

Spend 8hrs a day to work and 16minutes after the work to think on what needs to be done and/or the improvements that can be put in practice in the work that has been done in the last 8hrs. This 8hr-16mnts thought process will address the needs and requirements of improvements of day-to-day work initially and extend the same to the department and the company over a period of time.

From Here To Where: 'From Here To Where' is the thought process to analyse the current state of the company or institution, and the required next course of action. 'From Here To Where' thought process must not be the ordinary annual-strategy meets which normally look at how to outperform the competition, but it must grow beyond merely looking at competition to transforming the market.

Similarly, if not exactly same was the thought process that Steve Jobs applied at Apple, it wouldn't have taken the company from Personal Computers to Laptops, from Ipods to Itunes, from mobile phones to smart phones and Ipads, which transformed the entire computer industry, music industry and smart phone industry. His idea was never to outperform the competition but to think next of what new solutions to bring into the market. This kept Apple grow leaps and bounds to become most valuable company in the world and allowed Steve Jobs to transform the industry.

Need Statement:

After identifying the need, a need statement shall be phrased. Need statement is providing words to the thoughts. Words or jargons make concepts and ideas simple and clear like cross training, cutting edge, level playing field, low-hanging fruit, etc. The need statement shall be simple with minimum words and it encompasses the whole concept without needing an explanation. Need statement is very important as it; itself can do half of the convincing. Need statements can be like the following: Quick last-mile mobility; Green fuels for eco-friendly energy solutions, etc.

Examine the existing set up:

If we need to achieve two times the turnover for the company, we can look at organic growth like expanding into other markets or even the inorganic growth like acquiring the companies; if we need to have the quick last-mile mobility we have to go beyond widening roads and building flyovers.

Examine the need if it can be met with the existing setup. If so, assign the task to the executive or headship. Leadership is required to meet the need that cannot be met with the existing set up.

Conceive the task:

Conceiving the task is different from identifying the need. A need requires a solution. The 'task' is providing and implementation of the solution. For example, controlling the pollution in New Delhi is a need and odd-even programme is the task to meet the need. We all are very clear on the need to control the climate change, rein in the terrorism and alleviate the poverty. But how do we do that? Well, we do not know yet. We need to have a task for this purpose which is nothing less than a leadership task. Task is the course of action required to meet the need or improvement. It can be one task or many tasks that may be required to meet the need/improvement. Conceiving the task to meet the need is very important. The following mechanisms will help. Sometimes, the processes of identifying the need/improvement and conceiving the task may overlap.

Imitate:

Studying similar situations faced by others elsewhere, can help obtain inputs to identify the task for our need. The task can be imitated by tailor-making and fine-tuning it to suit our requirements.

Take the example of the need to increase the corporate efficiency. On examining the other companies which addressed the issue effectively, we will understand that implementing the digitalisation can be one of the tasks. This task is the imitation of what others have done already. But, implementation of digitalisation which is hitherto not known in the company requires a strong leadership. The major ideas like this one will have huge resistance as they intend structural changes in the system and its working.

Dissection of the need:

Dissection of the need is analysing the flow of the need, root-cause that has raised the need, and the fall-over if the need is not met.

Flow of the need:

Climate change is a serious issue that mankind is facing today. There is an urgent need to check the climate change. Climate change leads to droughts, floods, natural calamities, and increased incidences of diseases like cancer, increase of the pollutants like CO, CO_2 and the other greenhouse gases in the atmosphere and decrease of the useful gases like oxygen. Climate change is a threat to the existence of the mankind, flora and fauna.

Root-cause that has raised the need:

Root-cause analysis is the analysis of what action has triggered the problem and the need. Root-cause analysis provides clues on the possible task that can meet the need. Let us analyse the root cause for climate change. The main reason for climate change is emission of greenhouse gases and deforestation to accommodate the developmental needs. The emission of the greenhouse gases is the result of changing lifestyle of the mankind and the luxuries and comforts it desires. Forests can help absorb the green house gases and increase the quantity of oxygen in the atmosphere. Forests help increase the rainfall and maintain the seasonal changes intact. The rainwater harvesting decreased due to the diminishing catchment and lake areas is leading to reduced groundwater levels, droughts and over- dependence on rains.

On the analysis of the flow of the need and root-cause that gives rise to the need, we can derive one task which is the urgent action on building the forests and lakes.

Fall-over analysis:

Fall-over analysis is the analysis of what happens if we do not meet the need. Most of the world economies regularly face the issue of inflation. There is no economy which has not suffered due to high inflation. What happens if we do not rein in inflation? People will not be able to buy due to high prices. The need is to control the inflation. There are so many tasks for the same like controlling the cost and prices, controlling the money available in the economy through controls on the banking system, etc.

The fall-over analysis of what happens if the inflation is not checked provides us the information that purchasing power of the people comes down. The task hence, can be to increase the purchasing power of the people. This can be done through creating opportunities for earning higher income, providing employment opportunities for the unemployed or underemployed. This is possible through development and increased activity in manufacturing and services sectors.

Innovation centre

Innovation centres should be more of a serious activity with a better structure and increased emphasis. Teams shall meet regularly and discuss and members shall be able express their opinions without any hindrance and inhibitions. There shall be an innovation department in the organisation, whose function is to facilitate identification of the leadership tasks.

Corporate ecosystem

Corporates should encourage the employees to come out on their own with leadership tasks and allow them to do them. Corporates need to develop intellectuals in each domain of the tasks rather than having human beings working like machines.

Task Labelling

Corporates can create task domains based on the nature of the tasks like leadership tasks, headship tasks, executive tasks, etc. Corporates need to create a culture that differentiates leadership tasks, headship tasks, and executive tasks and ensure that there is no mix up of these tasks. By verbalising clearly what is the meaning of leadership tasks, a corporate can dispel the false notions of the employees about leadership.

Effectiveness of the task

Proving the task on paper is very effective to convince for going ahead. Effectiveness of the task is the on-paper evaluation of the various alternatives and establishing the better part of the proposed task. Take, for example, the need of improving the air quality in Delhi. Air quality in Delhi has deteriorated over a period of years. Particulate matter in the air has increased. Incidents of diseases have increased. Life expectancy of Delhiites has come down. The pollution in Delhi is due to vehicular emissions, emissions from factories and due to construction works. Number of vehicles on the roads of Delhi has increased by many folds over the period of last 6 to 7 years. Heavy dust rises from the ground due to vehicular movement and various heavy construction works. The peculiarity of Delhi is that the air will not get flushed fast, unlike the cities like Mumbai where the sea breeze flushes the city air regularly.

With dissection of the need process to identify the task to meet the need, we can clearly see different tasks that can meet the need. These are like reducing number of vehicles on Delhi roads, reducing factory emissions, reducing emission per vehicle, cementing the roads and sweeping the roads regularly to reduce the levels of rising dust, etc.

Banning or controlling the sales of vehicles to reduce the number of vehicles has direct impact on the economy, lifestyle and luxury of the people. Reducing emissions per vehicle is a long-drawn process that takes some years. But the need is urgent. The tasks like banning the older vehicles, restricting the heavy vehicles' entry, green tax, improving the pollution-free mass transportation systems like metro rail, etc. can be different short-term tasks. Odd-Even is another innovative task attempting to reduce the number of vehicles on the roads without actually doing the banning work.

Before taking on the implementation, proving on paper, the effectiveness of the Odd-Even – like showing the percentage of reduction of the vehicles on the roads and corresponding expected reduction of emissions–was important for the government and various agencies to engage all the stake holders and make them inclined to the task and dispel certain preliminary apprehensions.

Test the task

Google released Google glass, the augmented reality headset, in the year 2014. Google Glass is a wearable, voice-controlled Android device that resembles a pair of eyeglasses and displays information directly in the user's field of vision. Google Glass offers an augmented reality experience by using visual, audio and location-based inputs to provide relevant information. For example, upon entering an airport, a user could automatically receive flight status information. Users can also control the device manually through voice commands and a touchpad located on its frame.

The Google Glass operating system is based on a version of Android, and it can run apps called Glassware that are optimized for the device. The glasses have built-in Wi-Fi and Bluetooth connectivity and a camera for taking photographs and videos.

Certainly, as a tech-based product, it was the most promising product but it failed miserably. It was not well-received due to its style, privacy and functionality concerns. Google had to withdraw the product within short period after not so favourable response and reviews; it was a product launched and withdrawn hurriedly, even without conducting proper testing (However, Google is planning to re-launch the product with some more advanced features).

Testing of the task before its actual execution is essential. It is important to know beforehand whether the task delivers. Testing is required to avoid failure after the complete implementation, thus avoiding waste of efforts, time, money and other resources. It helps to fail fast and fail cheap.

What to Test

Before deciding to test the tasks, we need to understand the uncertain parts of the task. It is essential to test the only uncertain parts of the task. For example, in the task of building forests and lakes to meet the need of controlling the climate change, the uncertain part is the participation of the public in providing the land. It is this part of the task which is to be tested.

Different Testing Methods:

There are different testing methods available. However, selecting the appropriate one to test your task is most important.

Survey/Feedback

Survey/Feedback is one of the testing tools for the feasibility of the task. For the need to improve the top line of a company, one of the tasks can be to expand the overseas market. Market survey will help knowing beforehand, the effectiveness of the task and its feasibility.

Pilot Scale Implementation

The odd-even scheme in Delhi was implemented on a pilot scale for 15 days for its practicability, effectiveness, public acceptance and participation, etc.

Sample

Tasks like creating a new product or creating a new feature in the existing product can be tested by releasing the samples. This kind of testing is adapted by electronic or software companies for their new products. Samples can be released when the production cost of the sample is low.

Model

Model is made to represent or explain the concept or looks, like the windmill concept, advantages of direct drive over belt drive, architecture or design of a building, its colour scheme and arrangements, city plan, etc. Model helps in visualising. Model is made with completely different materials. Model can be a design on paper or in the computer. A model can represent one or a few or all the features.

Prototyping:

Prototyping helps testing the working. Prototype is not the small scale version of the product. Prototype is built to test the actual working of one or more features, not necessarily all the features. One of the tasks for the need to improve the top line of the company is release of the new products. Prototyping of the products will help test the product for consumer acceptance.

Experiment:

Experiment is a complete implementation of task or idea from end to end, in order to validate or refute the results. Experiments are conducted mostly for science theory testing, medicinal testing, etc. Experiment is the costliest of all the testing methods.

Once the testing is complete, the leadership task is ready for execution.

Execution of the Leadership Task:

Execution is not easy, even in spite of having a clear-cut task identified with the help of above framework. The questions like–What (exactly) to do? Why doing it? How to do it? Shall I do it? Can I do it? Why only I should do?–keep lingering. It requires leadership to cross these barriers and hence, there is a requirement of 'Leadership Engineering'.

Mark Zuckerberg created Facebook and revolutionised the social networking. In fact, creation has been his habit since childhood. According to writer Jose Antonio Vargas, "Some kids played computer games. Mark created them". He built a software program called 'ZuckNet' that allowed all the computers between his house and dental office of his father to communicate with each other. It is considered a 'primitive' version of AOL's Instant Messenger, which came out the following year. During Zuckerberg's high school years, under the company name of Intelligent Media Group, he built a music player called the Synapse Media Player that used machine learning to learn the user's listening habits.

Before Facebook, there were ideas like the photo address book which the students of Philips Exeter Academy–from which Mark had graduated- called 'The Facebook'. Such photo directories were an important part of the student social experience at many private schools. With them, students were able to list attributes such as their class years, their friends, and their telephone numbers. Earlier than Zuckerberg, three of his seniors at Harvard; Cameron Winklevoss, Tyler Winklevoss, and Divya Narendra were working on a social network called HarvardConnection.com. They had sought the help of Zuckerberg to build the network.

Zuckerberg's execution was perfect. He launched the Facebook. It became very popular in the Harvard campus. He decided to spread it to other university campuses like Columbia, New York University, Stanford, Dartmouth, Cornell, Penn, Brown, and Yale. On his mission to great things, Mark dropped out from Harvard and moved to Palo Alto, California. He hired office space, got investors and gradually scaled up the Facebook. Eventually, Facebook grew worldwide and by 2012, the number of users of Facebook crossed one billion.

The offers by major companies to buy Facebook were turned down as such offers were not in line with the idea–in the own words of Zuckerberg–of creating an open information flow for people.

The ideas similar to Facebook were already available before the creation of the Facebook. But had it not been via Zuckerberg's execution, the Facebook would not have been what it is today and there would not have been a revolutionary step in the area of social networking.

It is important to have the idea, know the need, devise and test the task to have the leadership task cut out. However, as explained in the previous chapter 'Decoding the Leadership Process', it requires Leadership Process to overcome the leadership barriers to take the task to execution. In the next chapter 'Engineering a Leader', it will be explained how to engineer a leader with the help of Leadership Engineering Framework to complete the leadership.

Framework for Identification of Leaderees:

It is required to have a leader to finish the leadership tasks that are identified with the help of the framework for identification of leadership task. Companies shall identify the leaderees and engineer them into leaders to have the leadership tasks completed. The following framework is suggested to identify the leaderee to subject him/her to Leadership Engineering-

- One of the affected
- Have volunteers
- Based on performance
- Subject to Leadership Readiness Tests (LRT)

One of the affected:

The leaderee shall not be one from the non-playing. The leaderee shall be from the middle of the action. The leaderee selected for the task shall be one who is affected by the need and who has suffered due to not meeting of the needs.

Take, for example, the social leaders like Nelson Mandela, Martin Luther King, Mahatma Gandhi, etc. They all rose from the middle of the action. They all had suffered by the lack of the cause what they fought for. They were not imported from somewhere else to take up the cause.

Have volunteers:

Somebody volunteers if he/she has great interest in the task. Asking for volunteers is a very effective way of identifying the leaderees.

Based on performance:

Have a shortlisted pool based on the past performance and feedbacks from the HODs.

Subject to Leadership Readiness Tests:

Leadership readiness tests are basically those that test the basic interest in the task, aspiration in the candidate and endurance against basic Dis-CATs- which are as explained below.

Test the basic Interest:

Call for proposals. Put proposals in abeyance with a reply that "You may contact us anytime". Check for follow-ups.

Test basic Aspiration:

Interest makes leaderee to start the task and see its preliminary success. It is the aspiration in the leaderee that takes him further. It is important to test the aspiration of the leaderee to pick him for the leadership engineering. The leaderee shall be subjected to leadership readiness interview. The interview can be generic or task-specific with questions like, Where do you see the task and yourself after three years? Why do you think you should do this task or do you think you can complete the task?

Test the endurance against basic Dis-CATs-

Brief the task to the leaderee, and devise a thread task. Create situations of discomforts, disappointments and distractions like sending the person on the test task, away from family for a period to a location which is not very comfortable.

ENGINEERING A LEADER

Leadership should be engineered and nurtured in an individual, in a corporate, in an institution or in a country. Companies, institutions or countries shall not wait for some leader to emerge on his or her own. Instead, they must engineer leaders and leadership proactively. Leadership can be engineered like any technology or product that a corporate or institutions develop. Some may have doubt, if it is possible to engineer a leader at all. The answer is 'Yes'. Engineering a leader is possible when it is taken up in a structured way.

Leadership in an individual or in a corporate will be hidden in the form of either a dormant leadership or a stifled leadership. This is because of the leadership barriers as explained in the previous sections. With the 'Leadership Engineering', the leadership, which is dormant or stifled due to the leadership barriers, is uncovered and put to use.

Leadership Engineering basically, is making the leaderee acquire the 'Leadership Vehicles' to move from one 'Leadership State' to the next by crossing the 'Leadership Barriers' and complete his leadership process. The leadership states, leadership barriers and leadership vehicles have been discussed in the earlier chapter of 'Decoding Leadership Process'.

In this chapter, a framework called 'Leadership Engineering Framework' is suggested to provide a structure to have leaderee acquire the leadership vehicles and complete his leadership process.

Leadership Engineering Framework

The Leadership Engineering framework essentially contains two areas with first one depicting the Leadership Vehicles and the second one depicting the Trigger Elements for the leadership vehicles, as shown below.

Leadership Engineering Framework						
Identify Task	Trigger wish	Trigger interest	Trigger aspiration	Trigger clarity	Trigger passion	Trigger urge

Leadership Vehicles area contains the following leadership vehicles-

- Identify the Task
- Trigger Wish
- Trigger Interest
- Trigger Aspiration
- Trigger Clarity
- Trigger Passion
- Trigger Urge

Each Leadership Vehicle in the Leadership Engineering Framework is associated with Trigger Elements. Let us study each element of Leadership Engineering Framework in the below sections-

Identification of leadership task

The first element in the leadership engineering framework is about identification of task. In the leadership process, the maxim is “Task First, Others Next”.

Leadership Engineering Framework						
Identify Task	Trigger wish	Trigger interest	Trigger aspiration	Trigger clarity	Trigger passion	Trigger urge
N-E-T-E-T						

This subject selection of leadership task has been dwelled upon at length in the previous chapter of ‘Identification of leadership tasks and leaderees’. Engineering a leader starts after the selection of the task. Let us now proceed to the other vehicles including their trigger elements.

Trigger Wish

Anyone born in this world has a wish. A wish is generated by seeing. Wishes are like–I want to be like him, I want to own a car, I want to have good food, house, etc. Anybody could have the wishes as having a wish does not cost anything. Having a wish enables having a momentary happiness. New wishes generate at every stage of life. Wishes are the starting points.

There is no need to put in any special effort to create a wish in any person. A briefing and presenting the benefits of the specific leadership task in the subject is sufficient to create a wish in the leaderee to take up the task.

Leadership Engineering Framework						
Identify Task	Trigger wish	Trigger interest	Trigger aspiration	Trigger clarity	Trigger passion	Trigger urge
N-E-T-E-T	Benefits of the Task					

The leaderee may stop at the wishing stage itself when he does not know what to do to fulfil his wish. This happens if the leaderee is highly ignorant. In these situations, the leaderee limits himself to 'Dream State'.

Trigger Interest

Ignorance stops a leaderee from moving from dreams to action. It is worth mentioning again, even if it amounts to repetition of what was discussed in the previous section–that the ignorance in this context is the ignorance of the direction and purpose. This ignorance is different from lack of knowledge or lack of awareness. A person may be highly learned or rich but may still look to be groping for something. This is because of the ignorance. A person does not become a leader if he is ignorant. He lives in his own world and does not know any other world.

Interest enables the leaderee to overcome the ignorance of the direction. Interest is required to generate the thirst in the leaderee. Vladimir Ilyich Ulyanov, alias Lenin was a Russian communist revolutionary. He was the architect of United Socialist Soviet Union. Widely considered as one of the most significant and influential figures of the 20th century, Lenin became an ideological figurehead behind Marxism-Leninism and thus, a prominent influence over the international communist movement. A controversial and highly divisive individual, Lenin is viewed by Marxist-Leninists, as a champion of socialism and the working classes.

Lenin was born to a wealthy middle-class family. Despite the emotional trauma of his father's death when he was just 16 years old and of his brother's death the next year, Lenin continued studying, graduated with a gold medal for exceptional performance, and decided to study law. His brother's execution turned Lenin to be interested in revolutionary politics. He pursued this interest risking academics and continued in this line of interest and became a revolutionary leader.

Similarly, each leader has got interest(s) triggered in his life. There have been incidents that have triggered interest in their lives. These incidents have happened naturally in their lives.

Leadership engineering framework attempts to trigger the interest in the leaderee without waiting for it to happen on its own. To understand what creates interest in a person, let us explore the interest enablers-

- Surrounding environment
- Family culture

Leadership Engineering Framework						
Identify Task	Trigger wish	**Trigger interest**	Trigger aspiration	Trigger clarity	Trigger passion	Trigger urge
N-E-T-E-T	Benefits of the Task	Surrounding Environment Family Culture				

Surrounding environment

Cricket is much talked about as a game in India. Cricket is present everywhere–in the TV, media, neighbourhood, classrooms, offices, club rooms, etc. The presence of cricket everywhere has triggered a lot of interest in the public of India, for cricket. This has led to success of cricket in India and produced many successful players. It led creation of most innovative league called IPL. Creation of IPL gave a new lease of life to the aspiring players, investors and business community.

Silicon Valley is a home to many of the world's largest high-tech corporations and thousands of start-up companies. It also became a global synonym for leading high-tech research and enterprises, and thus inspired similar named locations, as well as research parks and technology centres with a comparable structure all around the world.

Silicon Valley is a leading hub and start-up ecosystem for high-tech innovation and development, accounting for one-third of all of the venture capital investment in the United States. It was in the Valley that the silicon-based integrated circuit, the microprocessor, and the microcomputer, among other key technologies, were developed. As of 2013, the region employed about a quarter of a million information technology workers.

It started with the roots in telegraph, radio, commercial and military technology, blossomed by activities like Ham radio, U.S State funding of Fairchild's project, contributions by Stanford University, its affiliates, and graduates with the works–like that of Lee De Forest with his invention of a pioneering vacuum tube called the Audion and the oscilloscopes of Hewlett-Packard–Frederick Terman's (Stanford's dean) encouragement to faculty and graduates to start their own companies (he is credited with nurturing Hewlett-Packard, Varian Associates, and other high-tech firms, and is often called 'the father of Silicon Valley'), the works of William Shockley, the creator of the transistor and founder of Shockley Semiconductor Laboratory; Silicon valley provided a perfect environment for the tech companies to rise and flourish.

Encouraging elements, adding to the flavour, are the developments like -solid-state technology research and development at Stanford University, with the support from private corporations like Bell Telephone Laboratories, Shockley Semiconductor, Fairchild Semiconductor, Xerox PARC, and Terman's founding of venture capital for civilian technology start-ups.

These were followed by the silicon transistor, Fairchild Semiconductor, founding of Intel, development of computer networking, venture capital firms, software boom, launching of the ARPANET (precursor to the Internet), development of object-oriented programming, graphical user interfaces (GUIs), Ethernet, PostScript, and laser printers by Xerox's Palo Alto Research Centre (PARC); 3Com and Adobe Systems, Cisco, Apple Computer, Microsoft, Internet, dot com and so forth.

The environment that Silicon Valley has created and nurtured has encouraged interest in research, developing new technologies and starting new companies. More and more companies got created interested by the surrounding activities.

Family culture:

Why do we see more doctors from the family of doctors? Why do we see more business people from the family of business people? Why do we see some more singers from the family of singers? It is again, the presence of these factors in and around the family, which creates interest.

Interest Triggers

Taking cue from these interest enablers, there shall be an environment created around the subject in corporates, institutions, or organisations, which reminds the leaderee of the direction or task decided in each move. There shall be constant reminders.

Posters:

I know this company which I visited regularly as part of my previous assignments. Various posters were displayed all over the office in this company on the importance of the not wasting the food. Initially, employees of the office gave only a casual look at the posters. On continuously looking at the poster every day, there was a thought generated in the employees. The thought led to an interest after six months of watching the posters. The interest slowly resulted in practicing the food saving measures which were measured by the total weight of the food left out in the plates of the employees. After six months, the weight of the left out food reduced drastically to 60% of the initial amount.

Posters like these act as constant reminders and create interest in the beholders on the subject.

Publicity in the internal media:

Publicity along with a properly designed publicity material is another effective tool. The corporates should use their internal media like emails, newsletters, broadcasts, blogs, etc. for this purpose.

Repeated reminders:

Repeated communications on the task shall be sent to the leaderees. These continuous communications will act as reminders and make them have repeated thinking.

Encourage participation:

Frequent workshops, brainstorming, debates, etc. should be conducted in the organisation and the leaderees shall be made part of the participations in these events.

Trigger Aspiration

After the initial interest, there is a high risk that the leaderee will be prone to indifference. Due to the barrier of indifference, the interest is not taken forward. Indifference basically stems from the questions like–Why should I do? Or, Why I should do?

Indifference creeps in due to the factors like lethargy, excessive comfort, certain incidents in life, setbacks, no motivation, etc. Following are the trigger elements which trigger 'Aspirations' in the leaderee.

Leadership Engineering Framework						
Identify Task	Trigger wish	Trigger interest	**Trigger aspiration**	Trigger clarity	Trigger passion	Trigger urge
N-E-T-E-T	Benefits of the Task	Surrounding Environment	**Y-Factor**			
		Family Culture	**Social Status**			
			Incentives			
			Inspiration			

The Y- factor:

In one of the recent seminars I attended, one of the participants had a question after more than two hours of presentation by an eminent speaker. The questioner was at a senior management position in his company. He lamented that it was very difficult to get the work done as required, from some of his team members. His question to the speaker was how to motivate such people to have the results delivered?

The answer for such questions is very simple. We cannot motivate a person. The best way to address the indifference of the people, as mentioned by the questioner, is by explaining them the 'Why'. Most of the times, most of us work; whatever the work, we do it without knowing the 'why'. Most of the team leaders take all the pain to explain how and what needs to be done by their reports. But, the most important thing to explain is that why it needs to be done. The answer to the 'why' makes one have the aspiration to complete.

However, explaining 'why' is not very simple. The answer to

‘why’ shall be aligned to the needs of the person or desire of recognition in the person or to the benefits derived. It requires thorough understanding of the leaderee. The answer to ‘why’ shall be stronger than the efforts required for the task. The answer to ‘why’ is not a one-time activity. It requires regular reinforcement. If one is clear on ‘why’, he himself will figure out the ‘how’. The more we explain ‘why’, the less we would have to explain the ‘how’.

There are always big ‘Whys’ and big ‘Whats’; and small ‘Whys’ and small ‘Whats’ in a mission. The big ‘Whys’ and big ‘Whats’ are at panoramic level. Big ‘Whys’ shall lead to big ‘Whats’, which are again split into small ‘Whats’ and furthermore, explained by small ‘Whys’. For example, at corporate level, the big ‘Why’ explains the purpose of the corporate, which may be wealth creation for the stake holders or servicing the society by meeting its needs and the big ‘What’ is the product or service it chooses to fulfil the purposes. The big ‘What’ is split into small ‘Whats’ like choosing business model or a market. These small ‘Whats’ are explained with the small ‘Whys’ to have the required involvement and contribution from the executive.

Attraction of the Social Status:

Opportunities of promising social recognition, limelight in the circles shall be provided to the leaderee as part of the trigger to aspiration.

Incentives:

The attraction of the incentives is one of the main sources of aspiration. Incentives shall be connected with the task. For example, do not incentivise a mountaineer for climbing the large and difficult mountains with cash awards. It is an easy and simple thing to do. Rather, the mountaineer should be incentivised by granting permission to climb another mountain or with the permission to live in the mountains by allotting land or supporting him/her by opening a training centre for other mountaineers to make a living out of it.

Inspiration:

The leaderee shall be provided with opportunities to interact with his role models or spend time regularly with high profile or successful people. Such meets, interactions or taking part in the events–which these people attend, provide inspiration to the leaderees.

Trigger clarity

Mahatma Gandhi was one of the greatest leaders that the 20th century world had. He was an Indian freedom fighter. His creation was not just an independent India; his creation was an India filled with peace- loving, secularism-believing, independence-seeking, united, patriotic, self-reliant, democracy-loving, believing in non-violence, and development-seeking ambience. He created this ambience in India that ensured freedom of India along with showing the path and principles on which the country would progress. This ambience spread all over the world. It is not just India, but the entire world which is benefitting from his creation.

Mahatma Gandhi returned to India from South Africa in the year 1915. He was not clear; what to do and how to secure independence in India. He undertook an extensive travelling across India. Through the travelling, he was basically trying to get acquainted with his country. It helped him gather a lot of information about his country. The information he gathered provided him all the clarity. He got clarity on what actually was required for his country. He got clarity that his country needed not just freedom. His country needed self-reliance and poverty alleviation. It was with this clarity that he addressed the issues of his countrymen in a better manner. He could strike the aim precisely. It was with this clarity that he was able to fight for the cause and not pitch the fight against any individual, institution or country. It was with this clarity that he became dear to his countrymen and to the British as well.

The leaderee needs to have an enormous clarity to cross the leadership barriers called chaos, ambiguity, doubts and fears. The following trigger clarity in a leaderee-

- Information/ as many inputs
- Larger picture
- Seeing it as certain
- Seeing with open eyes what you see with closed eyes
- Writing your own constitution.

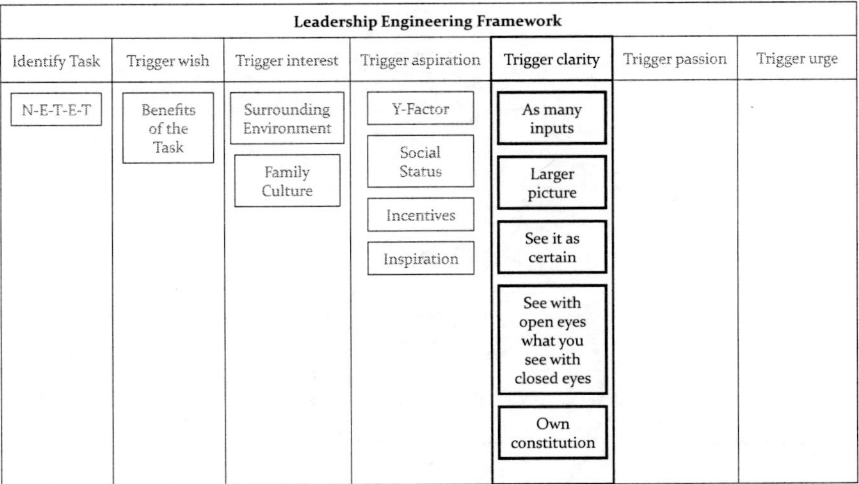

Information/as many inputs

Information provides clarity. Like the sun provides the light that guides us ahead in the real space, information provides the clarity that guides us ahead in the mind space. The more is the information one has the better is his clarity on the subject. To trigger clarity, the leaderee needs to be provided with as much information as possible on the subject he is working on.

Limited information leads to narrow perspective. It is the best fit here to remember the story of the blind people touching elephant as an example. It is a very popular story that requires no narration here. Each person in the story has his own perspective of elephant as they do not have any input other than what they touched. The person touching the trunk said it was like a snake. The person touching the legs said it was like tree. The person touching the body said it was like a rock. The person touching the tail said it was like a rope.

When all the inputs are collated and analysed one can understand that it is elephant that these people are talking about.

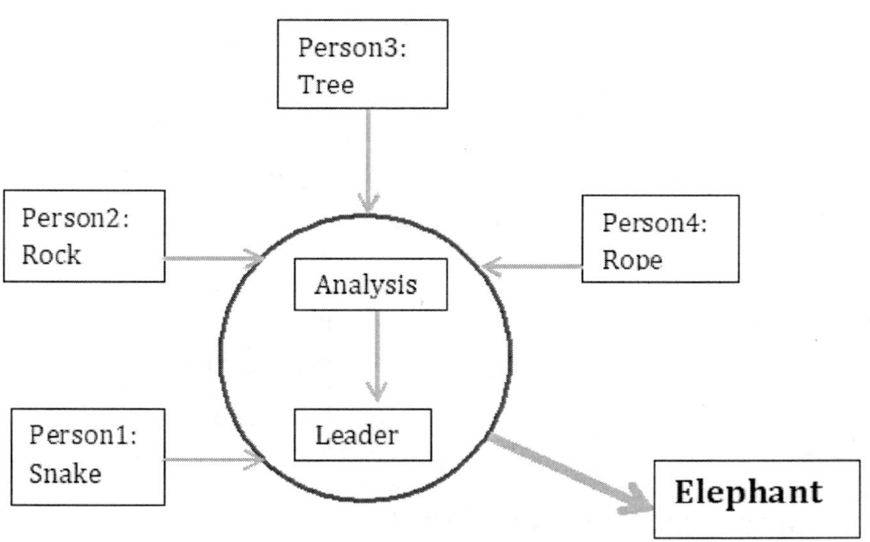

One or very few inputs lead to different understanding of the subject. Large number of inputs coupled with analysis lead us close to the actual picture.

The harm of narrow perspective with limited information:

Every person in the world has his own perspective on a given topic. The important thing is how close it is to reality and how useful it is to the objective. Any perspective framed on very little information is always a narrow perspective.

The young boy and the fate of the fish experiment:

The young boy and the fate of the fish story is a 5-part quick experiment that sums up the harm of the perspective framed on very little information. Given below are the five parts. Narrate each part to a separate group of people and ask them the question in the end.

The young boy is very hungry. He finds three live fish on the bank of the pond. What will he do with the fish?

The young boy is very hungry. He takes only vegetarian food. He hates the aquarium his friend has. His pet cat is hungry. He finds three live fish on the bank of the pond. What will he do with the fish?

The young boy requires fish oil massage for his skin disease. His pet cat is hungry. It hates fish in its food. He finds three live fish on the bank of the pond. What will he do with the fish?

The young boy requires fish oil massage for his skin disease. He always protects living beings. His friend has an aquarium. He finds three live fish on the bank of the pond. What will he do with the fish?

Each group gives different answers and there four different answers. One group answers the boy eats the fish. Second group answers the boy gives the fish to his cat as food. The third group answers the boy uses the fish for his skin disease. The fourth group answers the boy gives the fish to his friend for his aquarium. Each group depending on the information they have has a different perspective on the fate of the fish. But the actual fate of the fish is something else.

Narrate the entire story all parts together as below to another separate group and check the perspective they have now and how it is different from the others.

"The young boy is very hungry. He takes only vegetarian food. He requires fish oil massage for his skin disease. His pet cat is also hungry. It hates fish in its food. The young boy always protects living beings. His friend has an aquarium. The boy hates the aquarium his friend has. He finds three live fish on the bank of the pond. What will he do with the fish?"

The perspective formed by the group which heard the total story will be closer to reality. It is very important to gather as much information as possible and get as bigger a picture as possible–to frame a perspective for a very effective opinion and actionable plan.

Reading below the headlines:

Recently, I came across a newspaper article that had the headline saying "The trade deficit is nine-month low". The headline was very encouraging in the fact that trade deficit decreases when exports increase. I could have skipped reading further and could have gone to the next piece of news. But, I read further on the news only to learn that the exports in the month decreased. The chief reason for decrease in the trade deficit, however, was that the imports decreased more than the fall in the exports. The imports of the goods those were necessary but not available in the country, like crude oil, decreased which indicated the weak consumer sentiment and slowing down of the economy.

The headline, however, was true but gave a different perception from the reality which was realised only after the information was read completely.

Information and thus, clarity is obtained through readings, research, travelling, networking, and communication. The leaderee shall be subjected to trainings, brainstorming, exposure to bigger events in the company, interactions, visits, readings, etc. The process of providing information to trigger clarity should be followed consciously by the corporate, knowing very well what it is doing and why it is doing. Some companies may be doing some of these exercises, like trainings, without exactly knowing what they are doing this for. Being aware of the purpose in the act improves the effectiveness of the act.

See the entire/larger picture

By virtue, a leader sees the bigger picture. It is important in the leadership process that we see the bigger picture and do it as fast as possible. Bigger picture presents the holistic perspective; it helps take just decisions. It avoids framing narrow opinions. The action plans framed without knowing the bigger picture lead to a catastrophe.

The extensive travelling enabled Gandhiji to see the bigger picture. The other leaders hitherto were looking at only freedom from foreign rule. But due to the larger picture, Gandhi could understand, unlike many, that independence alone was not sufficient for India. With his larger perspective, he could address the real issues which made him different from other freedom fighters and endeared him to the masses.

How to get the bigger picture:

Talk to as many people/read/work/travel; take as many inputs as possible. Listen to as many opinions as possible. Professional networking helps getting the bigger picture. Small scale/pilot scale execution of the task provides bigger picture which is hidden. Testing the task, as discussed in the earlier section of framework for identification of the leadership tasks, helps leaderee get the bigger picture.

Deriving pattern from a chaos

Founded in 1998, Priceline is a travel-related website that helps users find discount rates and name their own prices on hotels, car rentals, airfares and vacation packages. Priceline shares jumped from $16 to $86.25 during its first day of trading in March, 1999, in the period of dot-com bubble.

The dot-com bubble (also known as the Internet bubble) was a historic, speculative bubble. The period of excessive speculation, mainly in the United States that occurred roughly from 1994 to 2000, was a period of extreme growth in the use and adoption of the Internet. Many investors were eager to invest, at any valuation, in any dot-com company, especially if it had one of the Internet-related prefixes or a ".com" suffix in its name. Venture capital was easy to raise. Between 1995 and 2000, the Nasdaq Composite stock market index rose 400%. It reached a price–earnings ratio of 200. In 1999, more than 19 large-cap stocks rose over 900% in value.

The dot com bubble, mainly due to the use of metrics that ignored cash flow and significantly overvalued stocks, eventually burst. The burst of the bubble, known as the dot-com crash, lasted from March, 2000, to October, 2002. During the crash, many online shopping companies, such as Pets.com, Webvan, and Boo.com, as well as communication companies, such as WorldCom, NorthPoint Communications and Global Crossing failed and shut down.

Several companies and their executives were accused or convicted of fraud for misusing shareholders' money, and the US Securities and Exchange Commission levied large fines against investment firms including Citigroup and Merrill Lynch for misleading investors. After venture capital was no longer available, many dot-com companies ran out of capital and went through liquidation. Supporting industries, such as advertising, travel and shipping, scaled back their operations as demand for services fell. Coupled with the increasing number of companies closing due to dot-com burst, the September 11, 2001 terrorist attacks crippled the entire travel industry.

The market conditions were highly chaotic for Priceline. In 2001-02, the share prices of Priceline plummeted to less than $10 nearly gutting the company. Most speculated that Priceline wouldn't survive. The sharp dip in the stock price should've buried Priceline entirely, but it managed to hold on.

In 2002, a new CEO, Jeffery Boyd, stepped in. Boyd's implemented a major shift in focus, putting emphasis on the quality of their service instead of looking at how it was performing in the stock market. Boyd accepted a $19 million loss in the beginning but slowly took the company forward to gain profits of $1.1 billion in 2011. The individual share price has gone from a mere $8 to $604. By 2011, the company was earning $4.8 billion as annual revenue and had a market value of about $29 billion.

Boyd did all of this by slowly eliminating all of the side business under the Name Your Own Price banner, like groceries, mortgages, insurance and even phone and new cars. He brought back travel and the services related to that. He also expanded beyond the Name Your Own Price system, introducing traditional methods of pricing as well. Because of this, the company is now being touted as one of the best success/recovery stories in the internet sphere. Priceline is now one of the leading online travel agencies in the world, re-emerging from failure to become a true industry leader.

Priceline clearly saw the pattern even in chaotic conditions; it showed that travel industry is evergreen and dot –com burst could be over ridden by focusing on the customer services and product line which is the inner lining for any company.

How to derive pattern from chaos:

Pattern in a chaos is some partial order which is constantly present from the beginning of the chaos till the current moment, obstructed by some crisscrossing elements.

In a traffic chaos, the traffic police will identify certain pattern in the traffic and those three or four vehicles obstructing the pattern, moving which the entire traffic can be cleared.

Looking for the following, carefully in a chaos, will help derive pattern from chaos-

- Elements or situations which are present/common for most of the time throughout the movement of chaos from the beginning of the chaos till the day
- The elements which are present before the chaos and missing in the chaos
- The elements which miss suddenly and reappear later in the situation of chaos
- Some partial order in the movement of chaotic situation till the current day from the last day or last week or last month or last year

See it as certain

We attempt only if the outcome is certain. Leaders see this certainty very clearly. They are more than confident of the outcome. They act. They act with vigour and passion.

Seeing the certainty is not very easy in this ambiguous, chaotic world. We are not sure what is going to happen to us tomorrow. Leader with his clarity can see the certainty. Leader sees the certainty of the result because he wants it. He is confident of it. He is passionate about it. He is determined that he will reach there come what may. His resolve is great.

What makes it appear certain? Analysing the below factors and learning from them will provide enormous clarity on the certainty.

1. Initial success/one encouraging incident
2. Failure Incident Analysis
3. If once done many times done

Initial Success/One encouraging incident

Larry Page created Google. The value of this creation requires no elaboration. It revolutionised the information flow. Larry Page along with Sergey Brin developed a page rank algorithm and used it to build a search engine. They implemented the idea and built a machine using their dormitory rooms in the Stanford University that connected their search engine with Stanford's broad band campus network. The popularity of the search engine grew among the Stanford users. Encouraged by the initial success, the search engine was made available to internet users and the mammoth Google got created.

The initial success made Larry Page see the certainty ahead clearly and he went ahead.

Failure Incident Analysis (FIA):

It is popularly said "FAIL is First Attempt In Leadership". There is no leader who has not failed. Failure is only in the means. But failure gives us the impression that the possibility is certain. Success is an incident that ends a process and triggers (shall trigger) another process. Failure continues the process till it is ended by success or is stopped abruptly. Success invariably, is followed by celebration. But failure shall be followed by analysis. Failure leaves an impression that had I done it the other way, I would have succeeded. That is why it is expressed that failure is the stepping stone for success. Learning from the failure takes the process to a quick end, which is success. Otherwise, the process takes longer or may be left without taking it further. Whether we attempt for the second time or not, whether we put the failure incident analysis to practice or not, is a different matter. Seeing the possibility, most of us do it again. Those who do not attempt again, they do not do so for the barrier of fear.

One is required to have the incident analysis done after the failure. This will show for certain that the success is certain. Look at those leaders who failed initially and had great success later on, like Soichiro Honda (turned down by Toyota for a job but ended up creating Honda), Akio Morita (never got discouraged by the failure of his first product rice cooker and went on to build Sony), Abraham Lincoln (had numerous failures in the life but went on to be one of the greatest presidents of USA), etc.

The failures provided inputs to these leaders and helped them see the possibility of success clearly.

If once done many times done:

Mount Everest is the tallest peak in world with a height of 29029 ft. The first expedition to set foot on the summit started in the year 1921. But it took 32 years for first success that came through Edmund Hillary and Tenzing Norgay in the year 1953. And there has been no stopping since then. Mountaineers have been summiting the peak every year. Over 4000 people have climbed the mountain since then. Tenzing and Edmund showed the world it is possible to climb the peak. They provided the clarity on the possibility and many followed. Peaking the summit has been made much easier. The route has been standardised with camps set up at important points. Service providers sprung up to guide the professional mountaineers and enthusiasts. There are people who climbed the summit more than 10 times. Disabled people have climbed. People have skied down the peak for some distance. The peak has been climbed even in the most horrifying winters. Girls as young as 12 years have climbed the summit too.

The climbing of the Mount Everest by the first people provided enormous clarity for the rest. Many did later on and people continue to do it. If once done, that brings clarity and it is done many times.

See with open eyes what you see with closed eyes:

One may think there is no question that we cannot see anything with open eyes and the opposite is not possible. The paradox is it is difficult to see with open eyes what we see with closed eyes. Think or imagine something with closed eyes and it disappears the moment you open eyes. This is because the mind with closed eyes has no distractions. The moment we open eyes we get to see many things which immediately take us away from the thought we just had with closed eyes. This can be for example, a person next to you and the colour of the shirt he is wearing and you will immediately feel like saying "Wow, what a nice shirt", forgetting whatever thought you just had with closed eyes.

The person, who can see his thoughts, imaginations or plans vividly with open eyes wading through all the distractions of the bright light, attains highest clarity.

Write your own constitution:

Like the nations, individual also must have their own constitution which will guide them constantly on the purpose of their life; the principles, values and virtues they believe and follow. The constitution shall also have the milestones and timeframes stipulated to measure the progress. One should start writing his constitution from the age of eighteen. It is never too late even if it is written in the age of forties. Writing constitution makes the thoughts and ideas of the individual explicit. The constitution can be amended periodically and can have add-ons constantly.

Trigger passion

Passion is being on the task all the time. With the interest, aspiration and clarity, the leaderee is already half passionate. The fear of barriers like discomforts and disappointments hold back the leaderee. Passion is the inner fire that requires no fuelling once triggered. Passion is like reactions in the sun which last forever. Passion once triggered lasts. Following help trigger passion-

- Have a child to love a child
- Ownership
- One of the affected
- Saying No
- Overcoming False Justification
- Deep thinking
- Energy for the mind
- Familiarity with the dis-CATs

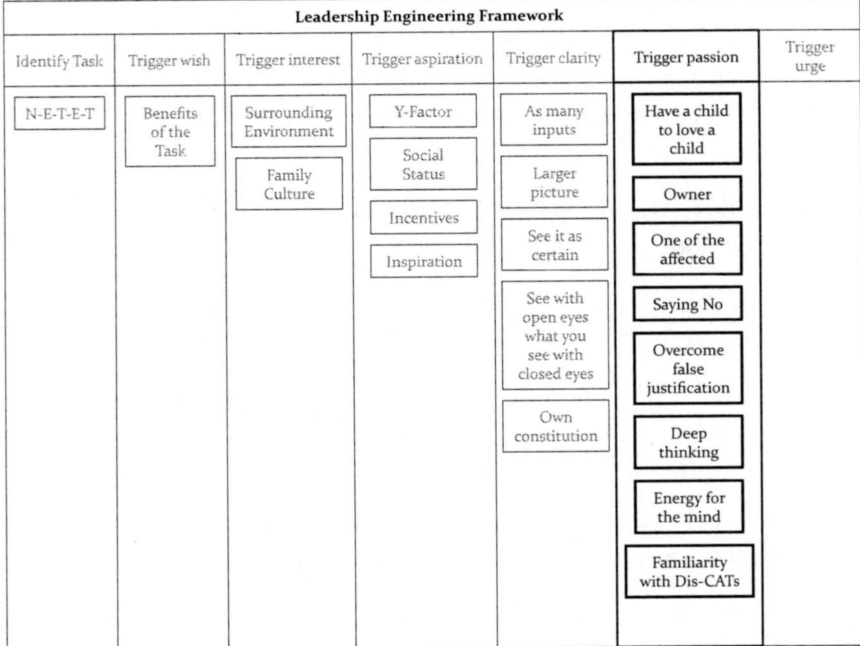

Have a child to love a child

You can love children only when you have them. One needs to have the task in hand to be passionate about. Having the task is the first step in being passionate. There shall be something to be passionate about. Isn't it?

Ownership

Create the sense of ownership in the leaderee by involving him/her in the entire process including decision making.

Overcoming False Justification:

Justification is the biggest art humans have. All the good done is justified and bad done is also equally justified. We can justify stealing, we can justify indulging in corruption, we can justify even killing. Terrorists are the biggest source to learn the art of justification. Many a time we get tempted to justify and give reasons for not doing what is required to do at that point in time. These are all false justifications. It is very important to the leaderee to know; which are true justifications and which are false ones.

The following are some of the examples of false justifications for the ready reckoning for the reader-

- I am under tremendous pressure from my boss for achieving the sales target. What if I deviate slightly from norms to achieve the target?
- Nobody is watching me and there is no way people will come to know of the wrong doing. So let me do it.
- Nobody has treaded this path/followed this model/taken this stance/taken this risk. So I will not.
- I am poor. So what is wrong in stealing once to survive?
- Everybody, including my bosses/my friends/the government officials, is siphoning the money. What is wrong if I do?
- I will compromise the moral only once. Just only once.
- Why only I must work when everyone seems to be enjoying?
- What if I do not complete the task today? There is whole lot of time in the life. Let me do it tomorrow.
- I am too busy to do this priority task today.
- Is it not the responsibility of all? Why only I shall be concerned?
- It is because of the faulty system/government/society. There is nothing I can do about it.

The leaderee shall undergo the consciousness process to overcome false justifications.

Saying No

Being passionate is saying "No" to most. Saying No helps weed out distractions. In order to cling to one, we have to say "No" to hundred.

Energy required for the mind

Mind requires energy like our body does, to do a task. Our body requires energy to move the body parts to do the task. Our mind requires energy to generate, move and firm up the thoughts seeking to do the task. Act of not doing what is required, comes mainly from lack of energy to spend on the mind to move the thoughts to do the act. We spend little energy for this purpose and hence brain moves towards whatever thought that prevails at the time. And we end up justifying the act.

Spending energy on the mind is very important. The following tip will help practicing spending energy to move the thoughts in the mind. Whenever you are unable to make your mind shift to the required thought you physically walk ten meters away from the point where you are standing/sitting/lying. Think about the thought you wanted to have. This thought emerges to the surface. Imagine you are physically moving the thought from the bottom of the stack to the top.

Deep thinking:

Deep thinking aided by Interest, Aspiration and Clarity will trigger Passion. Meditation helps for this purpose. Meditation is deep thinking on a subject keeping the mind from wavering. Deep thinking shall be carried out every day. There need not be any regular time for this purpose. Deep thinking can be practiced whenever the leaderee finds free time. It only requires initiating the deep thinking. With the help of already triggered interest, aspiration and clarity it becomes easy to sustain the deep thinking and trigger the passion.

Familiarity with the discomforts and disappointments

Dreadful imagination is more fearful than the real. Familiarity facilitates to come out of the imagination, and know the real. It increases the comfort.

Horror movies do exactly the same. They will make the viewer imagine. Once the suspense opens up, there is no more horror left. Send a group of people into a dark room and see the fear in them. Send the same people into the same dark room repeatedly and after going into the room for 3 or 4 times, the familiarity with the room increases and the fear is gone. When they go for the first time, they are not familiar, hence they fear.

Create small discomforts and disappointments in the leaderees in the task to make them be familiar with the discomforts and disappointments and never fear them.

Consciousness process of distractions

Consciousness process of the distractions is being aware of the distractions. Imagine and jot down the possible distractions. This enables to be conscious of the distractions when you are actually distracted. If you are aware that you are distracted or what can possibly distract you, you can come out of the distraction or escape the distraction relatively easily.

One of the affected

There is no more passionate person than the one who has suffered. An outsider cannot be passionate enough for the cause. Be involved. Be part of the need. Experience the suffering. Be in the middle of the action. One cannot be a non-playing captain and still win the match.

Trigger Urge

The leaderee falls short of being leader due to three main reasons; the fear of the leadership process due to the barriers involved, having no clear leadership task in mind or having not understood leadership.

The leaderee travelling this far in the leadership process and who has the leadership task cut out will have the urge automatically. However, the leadership barriers at this stage which are resistance and hardships may seem insurmountable at the initial stage of this State. The clear understanding of the following will help sustain the 'Urge' in the leaderee-

- Convincing to handle resistance
- Leader Need
- Leadership Compulsion
- Love for Subjects
- Understanding Leadership

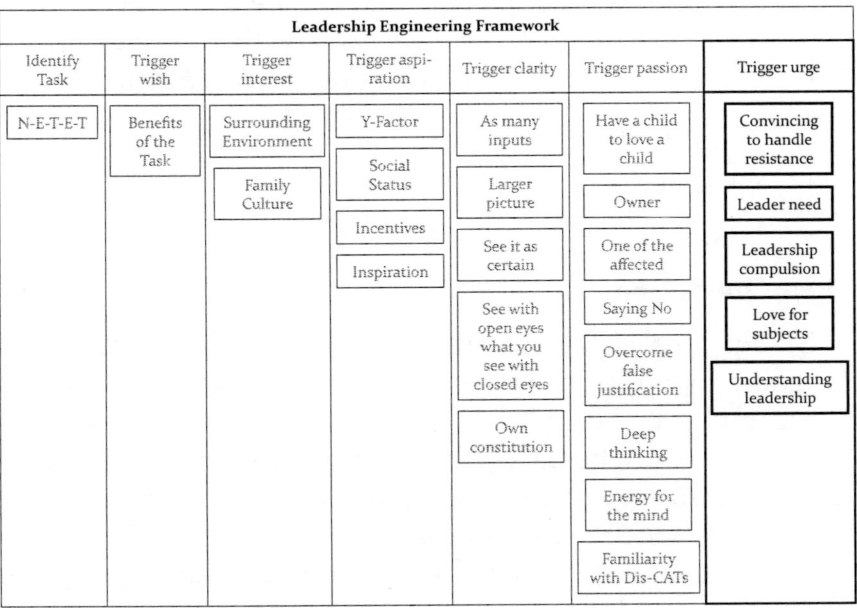

Resistance and Convincing:

As explained earlier, the first thing that a new idea says hello to is resistance. Resistance crops up at each stage of leadership continuously and repeatedly. Leader in the leadership process that is in the process of creation, does lot of convincing, proving and showing to overcome the resistance. Convincing is having agreement. Convincing has two parts- a) Getting convinced (convincing self) and b) Convincing (others). While clarity helps the leader stay convinced, convincing is equally difficult if not more. A leader spends most of his time on convincing. Leader is required to convince all the stake holders in the leadership process. These stake holders are called leadership subjects.

Leadership subjects

Leadership subjects are the stakeholders in the leadership process. Leadership deals essentially with three subjects, viz. Beneficiary, Accomplisher and Opposition (BAOs).

Leadership subjects are-

- Beneficiary
- Accomplisher
- Opposition

Leader works for the beneficiary, with the accomplisher against opposition. Beneficiary benefits from the leadership, accomplisher helps leader in the execution of the tasks and opposition can be those who are sceptical about leader/means/ tasks or who have conflict of interests. It is not necessary that the subjects, which are beneficiary, accomplisher and opposition, are three different groups. They can be the same group acting differently at different stages of leadership.

Steve Jobs was the co-founder, chairman, and chief executive officer of Apple Inc. and founder, chairman, and CEO of NeXT Inc. Jobs is widely recognized as a pioneer of the microcomputer revolution of the 1970s and 1980s. Jobs's leadership reaped benefits on Apple, stake holders of apple, board of directors and the customers. He hired an excellent team and along with the co-founder Steve Wozniak, they became part of the accomplishers of the Jobs's leadership. However, Jobs was not without opposition.

The greatest opposition that came to Jobs was from his own accomplishers and beneficiaries like the then CEO John Sculley, board of directors and stake holders who believed Jobs was crazy. Jobs was not allowed to do what he felt was right. His powers were curtailed and eventually, he had to leave the very company that he himself created.

Leadership styles are acts of convincing

Leader adopts different styles. These styles are different acts of convincing. On the analysis of different leadership styles proposed by Daniel Goleman like the coercive style, authoritative style, affiliative style, democratic style, pacesetting, coaching, etc. it can be noted that all these and many other different styles seek only one thing which is agreement from the subjects. The basic idea of any leadership style is convincing the subjects.

Elements in the flow of convincing

Convincing, thus, is an important part of leadership to overcome the resistance. There are different elements in the flow of convincing as illustrated below.

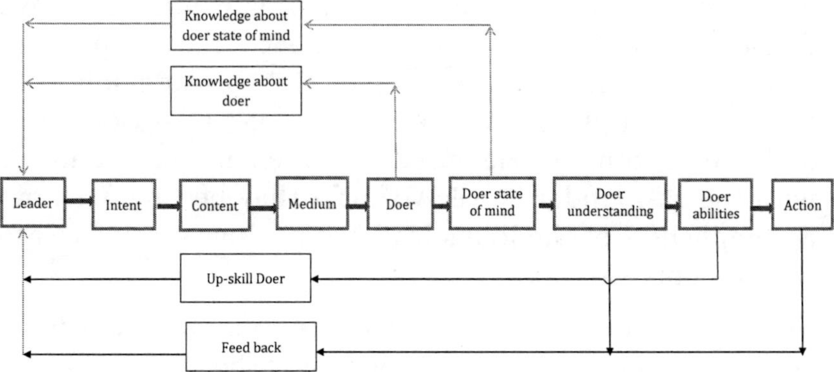

Convincing continues till the action falls in line with the intent

Leader starts with the intent. He acquires knowledge of the BAOs (Beneficiary, Accomplisher and Opposition) and their state of mind. He prepares the content and chooses medium for convincing BAOs. There will be action depending on the state of mind, understanding and abilities of the BAOs. Leader takes the feedback of the understanding of the BAOs and evaluates the action. If the action is not in line with the intent, the leader adjusts the content and medium which aligns the state of mind of the subjects, refines the understanding, improves the abilities which results into better action. The whole process is convincing.

Convincing continues till the actions fall in line with the intent, which may take many convincing cycles. The opposition and failures seem insurmountable in the early cycles. Leaders keep attempting till such time, exhibiting enormous patience and urge. Most of the leadership attempts fail as they do not pursue the cycle till the end or they change the intent due to lack of conviction.

Steve Jobs understood his subjects and their state of mind well. His content and medium was to withdraw from action and prove himself by doing it elsewhere. He founded NeXT, reshaped Pixar, underwent the entire struggle and faced all the hardships again. His fortune was squandered. He overcame the distraction of quitting and persisted. He kept on. Apple was finding it tough to sustain the momentum after the exit of Steve Jobs. Its sales decreased. Company's valuation was coming down. There was a growing desire in the subjects that he comes back and takes over the reins. His second term at Apple was now unopposed. He put the company back on track. The stocks grew from around $16 to more than $400.

The different elements in the flow of convincing are explained as below-

Subjects:

Subjects exhibit basically three types of behaviours-

Subordinate (child)

Peer (adult)

Superior (mature)

The leader feels convincing subordinates is easy. Subordinates have childlike behaviour. They tend to follow and obey the leader like a child to a parent or subordinate to a boss. Their behaviour is called subordinate or child behaviour.

Convincing peers is very difficult. Peers feel they know better. They feel they are right and others may be wrong. Their behaviour is adult like. They lack the listening. They act to compete. They tend to feel jealous and bring in their personal interests. Leader shall deal with the adult ego.

Some of the subjects the leader deals have mature behaviour. Their behaviour is like that of superiors to subordinates or grandparents to grandchildren. They behaviour is like counselling. They tend to listen and understand.

All these behaviours are experienced in all the three types

of subjects, irrespective of whether they are beneficiaries, accomplishers or opposition.

Intent:

The intent of leadership is the nature of creation; the creation that benefits. Leaders shall possess strong intent. In the entire cycle of convincing, the intent should not change. Convincing collapses and becomes unwarranted if the intent changes.

Content:

The actions of leader like skills, communications, strategies, literature, etc., forms part of the content.

Medium:

Medium is basically the channel of communication. Leaders communicate through actions, speeches, writings, meetings, etc. In the cycle of convincing, leaders adopt different media like public speeches, writings, actions like processions, marches, etc. like Dandi march of Mahatma Gandhi, processions by Martin Luther King, etc.

Subject's state of mind:

State of mind of the subjects is very important. Leaders need to understand the state of mind of the subjects very clearly in the cycle of convincing. For example, when the leader is trying to convince a terrorist of the current day nature, he shall understand that their mind is filled with hatred, revenge and anti-freedom feelings.

Subject's abilities:

The leader shall understand the subject's abilities and provide support to enhance these abilities to obtain the desired action.

Action:

Action is the last element and the objective of the convincing. The leader repeats the cycle of convincing till the actual action meets the desired one.

To change the course one need sync first:

It is highly difficult to push a speeding car back. The effective way is to get into the car and steer its direction gradually and till it completely turns back. This way the car turns back with the same speed without losing its momentum, unlike in the pushing where the car has to come to halt completely and move backwards.

You cannot change the course by being external to it. This is the very reason for most of the Promoter-CEO conflicts, like in the case of Cyrus Mistry and the Tatas. You need to sync first. Syncing will win you the acceptance. While syncing, the most important part to be mindful of is that-you shall be one of the course, but not part of the course.

Leader Need

Urge is the strong desire to fulfil a need. Leadership urge is a strong desire to fulfil *'leader need'*. The leader possesses a strong leadership characteristic need called leader need. The *leader need* is explained in the below.

We all agree that needs drive us. Needs make us do what we do. Needs can be categorised basically into six varieties, which are survival needs, physical needs, emotional needs, social needs, inner needs and contribution needs.

Survival needs, physical needs and emotional needs are primary needs. Social needs, inner needs, contribution needs are secondary needs which are met after meeting the primary needs. In each category, there are basic needs and there are enhancing needs. Basic needs are those that are must to be met. Enhancing needs are for the betterment after meeting the basic needs.

The needs take the precedence in the order of basic survival-enhancing survival, basic physical- enhancing physical, basic emotional-enhancing emotional. The secondary needs start with basic social needs only after fulfilling both basic and enhancing needs of the primary needs. There is no order of precedence in the secondary needs after the basic social needs.

The different needs in each category are tabulated as below.

	Primary needs			Secondary needs		
	survival Needs	Physical Needs	Emotional needs	Social Needs	Inner Needs	Contribution needs
Basic	Food	Comfort	Kindness	Appreciation	Knowing	Give
Enhancing	Clothe, Shelter	Pleasure, Fitness, Sex	Love	Recognition, Self Esteem, Status	Wisdom	Fulfil the needs of others (Create)

The *'leader need'* is the element in the dimension of enhancing contribution need which is the need to *"fulfil the needs of others".*

The urge in the leaderee turns this 'leader need' into the basic survival need for a leader and makes it a must to meet. Urge never allows leaderee to stop due to hardships. One who has the urge is never stoppable.

Leadership compulsion–Compulsion as strength

Why lion is stronger than the elephant; compulsion for survival. Why child is stronger than the parent; compulsive desires like eating chocolates.

I may like to narrate a small story at this point. Recently, when I was travelling to my hometown by train I witnessed a small yet powerful incident. As I was working on editing this book while travelling, I heard a small child crying continuously. The child wanted ice cream from the vendor who was passing by, calling loudly for selling. The young mother who was the lone traveller with the child was resisting buying. The child kept crying with increased volume. It caught the attention of the fellow travellers who kept casting the looks at the child. The mother was trying to be equally strong to not to yield to the boy. It is anybody's guess who won this skirmish. Such incidents happen in every parents and children life.

Anything sought as compulsion is achieved. Leaders make leadership as compulsion. The compulsion turns out to be their strength.

Love for the subjects

Love is offering what pleases right. Love is one side of the coin. The other side is worship. Worship is what we return to love. This is true for all forms of love. The love between mother and son (mother loves-son worships), the love between teacher and student (teacher loves-student worships), wife and husband (one loves, the other worships), leader and beneficiaries (leader loves and beneficiaries worship).

The urge in the leaderee comes from the love for his subjects. The love generates from interactions. Constant company, being one of them and networking with the subjects creates love. It is bound to occur in our mind whenever we interact with a person we get reminded about the negatives of the person in the mind. The leaderee shall not entertain having the negatives of others in the mind while dealing with them. This will make the leaderee have love on the subjects.

Understanding Leadership

Revisiting the understanding leadership deepens the urge. Leadership is the best purpose a life can have. Leadership is creation. Only leadership can provide the sustained solutions to the world problems. Leadership is the genesis for all the turning points in the development of mankind. Leadership is immortal and hence the leader. Leader is the most loved, remembered and admired.

Understanding leadership is the great source of Leadership urge.

The complete Leadership Engineering Framework:

The complete Leadership Engineering Framework containing Leadership Vehicle along with their trigger elements is presented as below.

Leadership Engineering Framework						
Identify Task	Trigger wish	Trigger interest	Trigger aspiration	Trigger clarity	Trigger passion	Trigger urge
N-E-T-E-T	Benefits of the Task	Surrounding Environment	Y-Factor	As many inputs	Have a child to love a child	Convincing to handle resistance
		Family Culture	Social Status	Larger picture	Owner	Leader need
			Incentives	See it as certain	One of the affected	Leadership compulsion
			Inspiration	See with open eyes what you see with closed eyes	Saying No	Love for subjects
				Own constitution	Overcome false justification	Understanding leadership
					Deep thinking	
					Energy for the mind	
					Familiarity with Dis-CATs	

How to use Leadership Engineering Framework:

The relationship of Leadership Engineering Framework with the leadership process is similar to that of a design process with a construction process. In the process of engineering the leader, each step taken shall be verified regularly with what it contains in the Framework. The point of deviations shall be observed and noted. Corrective actions shall be taken to fall again in line with the framework from the deviations, right at the point of deviation. Each leaderee who is being engineered shall have the framework always in mind and shall keep having the regular crosschecks. The simple techniques like having the printout always displayed at the work desk or always keeping it in the pocket will help a lot.

The leaderee shall be evaluated regularly to find out what leadership state he or she is at. What are the different leadership vehicles and their trigger elements the leaderee has used and what are the leadership barriers leaderee has crossed so far are to be noted and the outcome must be measured and compared with the desired one. The leaderee shall be assessed of his preparedness to move further from the current State. Leadership Engineering Framework shall be referred to find out and keep ready the Leadership Vehicles and the trigger elements that the leaderee requires to move further from the current state.

Leadership Engineering of own self

The Leadership Engineering of own self is different from doing that for a corporate talent. The interest of a corporate is to engineer leadership for the tasks that it is interested in. For the corporate, the tasks or the results expected are pre-specified and the leaderees are identified and engineered to meet the requirement.

However, the Leadership Engineering (LE) of own self is slightly different. For own, Leadership Engineering is not the engineering of leadership in the area of someone else's requirement but it is the engineering of leadership in his or her own area of interest. Leadership Engineering of own starts with identification of own interests followed by all other steps as discussed in the above sections. The only difference is that for a corporate leaderee, the corporate provides all the trainings and support for leadership engineering but for own, he/she has to undergo leadership engineering on his or her own.

There are different methods to check the area of interest of own self as mentioned below.

- Encouraging day dreams of own and tracking them
- Subjecting own self to psychometric tests
- Subjecting own self to "Find my Interests" activities

The interests of own self shall be known from the collective results of these methods. The rest is to follow 'Leadership Engineering Framework' as explained in the chapter.

LEADERSHIP HABITS

"Man forgets what he preaches while practicing and what he Practices while preaching"

Habits are actions performed constantly both consciously and unconsciously. Leadership habits are attitude, values, virtues, skill, actions and grip on leadership vehicles displayed constantly, both consciously and unconsciously.

However, leadership does not start after acquiring the habits. Leadership starts with identification of a leadership task. Leadership process makes the leader acquire the leadership habits. Leaderee acquires these habits in the due course of leadership even if they do not have these habits initially. The Leaderee uses these habits to accomplish the task.

Mahatma Gandhi's greatest innovations in his fight for independence of India were the Dandi march and the hunger strike. These innovations were the results of serious thought process on how to accomplish the already cut out task.

If you want to reach Sidney from Perth and you start walking, you will reach in 60 days, you take a cycle you will reach in 40 days, you take a car you will reach in 4 days, you take a train you will reach in 2 days and you take a chartered aircraft, you will reach in 5 hours. By any means, reaching is evident as you know the destination. Even if you have your own aircraft and start the journey without knowing the destination, you will be continuing to hover without reaching anywhere and eventually land back at the same place. Knowing where to go is the first essential part in the leadership process. Starting without knowing the destination is like moving but not going anywhere.

So do not focus on acquiring the habits in the very beginning (Habits include the skills which is explained in the below section). Focus first on what to do. It is a reverse approach to acquire skills and habits first and then deciding what to do. Most of the leadership training programmes attempt this reverse approach and lament for being ineffective. These trainings are best if they are provided to the people who have clear leadership task in mind. If you decide on what to do, you will know what skills you require to complete what you want to do and can focus on acquiring just those skills. Decide on the task first and acquire the skills required for the task is the beginning to end approach. As we have already discussed on what the leadership task is and how to select the leadership task in the earlier sections, let us now focus on Leadership habits.

Leadership habits make the leader distinct from others and yet the thrust is on the creation. No leadership (creation) sustains unless it is based on the leadership habits like right attitude, virtues, and right skills. It is famously said that one is respected for his "Thinking-Being-Knowing -Doing" that is, for his "(Thoughts)–(Virtues)–(Knowledge, Skills)–(Actions)".

Leadership habits are mainly of two fronts-

- 3CVA Qualities
- DoSo Skills

3CVA Qualities

3CVA Qualities make the most important part of the leadership habit system and constitute the main difference between the leader and the others. 3CVA Qualities are the qualities of Compassion, Contribution, Compliance, Virtues and Actions. The leader will posses these qualities and will exhibit them in all his deeds as it is the very nature of leadership, its definition and purpose. The following are the most important 3CVA qualities which are also are explained in a nutshell in the below table-

3CVA Thoughts				
Compassion	Contribution	Compliance	Virtues	Actions
Do Good	For Liberty	Truth	Aggression	I Can
Helping	For Prosperity	Unity	Confidence	I Shall
Respect Fellow Beings		Discipline	Emotional Intelligence	Now, Not Later
Be Grateful		Integrity	Focus	Decision Making (This Or That)
Gratitude		Honesty	More Actions-Less Words	
Trust		Responsibility	Passion	
Hurt Not Others		Shall Follow Guidelines	Patience	
Belittle Not Others		Shall Follow Law	Purity	
Appreciation		No Misuse	Self-Awareness	
Caring		No Taking Undue Advantage	Sincerity	
Loving		Not Being Greedy	Stretching Beyond The Comfort Zone	
Loyal		Not Being Jealous	Thought Intelligence	
Nice Words/ Actions		Not Doing Wrong	Willingness To Change	
Tolerance		Not Harming	Wisdom	
Smile		Not Stealing	Energetic	
Empowering		Shall Not Be Selfish	Fearless	
Obedience To Elders			Pleasing Demeanour	
Others First			Calm	
			Inner Peace	
			Humble	
			Trust Worthy	
			Willingness To Change	
			Wisdom	
			Confidence	

Qualities of Compassion:

Qualities like doing good, helping, respecting fellow beings, being grateful, trusting, not hurting others, not belittling others, appreciating, caring, loving, being loyal, saying nice words, doing nice actions, having tolerance, smiling, empowering, obeying elders, and putting others first are some of the qualities of compassion.

Qualities of contribution:

These qualities include contribution for the liberty and the prosperity of the mankind.

Qualities of Compliance:

Truth, unity, discipline, integrity, honesty, responsibility, following the guidelines, following law, not misusing, not taking undue advantages, not being greedy, not being jealous, not doing wrong, not harming, not stealing, not being selfish are some of the qualities of compliance.

Qualities of Virtues:

Most common virtues that form the 3CVA Qualities of leadership habit system are Aggression, Confidence, Emotional Intelligence, Focus, More actions Less words, Passion, Patience, Purity, Self-awareness, Sincerity, Stretching beyond the comfort zone, Thought Intelligence, Willingness to change, Wisdom, Being Energetic, Being Fearless, Pleasing demeanour, Calm, Inner Peace, Humble, Being Trustworthy, having wisdom and confidence.

Qualities of Actions:

I can, I shall, Now-Not Later, Decision making (this or that) are qualities for actions.

Thought Intelligence

Thought Intelligence is the most important part of the 3CVA Qualities and makes the most important leadership habit. It is famously said as "Watch your thoughts, they form your words;

Watch your words, they form your actions; Watch your actions they form your habits; Watch your habits, they form your character; watch your character, it forms your destination".

Thoughts are the basis for anything and everything. The merit of a person is determined by his thought process. The basic differentiator between two persons is their thought process. The thought process makes the person what he is; good or bad, rich or poor, successful or failed, beautiful or ugly, healthy or sick, ruler or a subject, happy or unhappy, calm or seethe, wise or fool, honest or thief.

Controlling thoughts controls everything. The Thought Intelligence helps achieve the feat. Thought intelligence in a nutshell is the process of knowing and controlling your thoughts. However, it requires an elaborate understanding to practice Thought Intelligence.

Let us start understanding Thought Intelligence by understanding the thoughts first. I have been for very long, undergoing the process of understanding thoughts. The process has led me to realise that I should understand the more basic subjects like Life, Soul, Mind and Body and then develop the understanding of thoughts on the foundation of the above understanding.

I have read a lot of books, listened to the lectures of many *dharmic* gurus on the subject. But none could give me the satisfactory explanation of what the Soul, Mind and Body are. I have been trying to reason out on the subject with the available knowledge and wisdom and happenings around in the real world. Soul, Mind and Body form the essential of life. The difference between living and non-living is having or not having all three of these.

Body is the simplest thing to understand in the sacred trio of Soul, Mind and Body that form life. Body is the physical structure of a living being that can be seen and felt. The human body consists of the heart and the circulatory system, the brain and the nervous system, the stomach and the digestion system, the muscular system that gives movement to the body, endocrine system that releases hormones, immune system that protects the body, reproductive system, skeletal system, lungs and respiratory system, the sensory system, etc.

Mind, on the other hand is the set of thoughts the living beings have. It is the mind that differentiates the primitive living being and the advanced living being like humans. The primitive living beings have only basic mind which includes the thoughts like eating the food and running for safety. More the sophisticated life form, more developed is the mind. Humans, the most developed of all the life forms, have mind that includes about more than the mere food and shelter. It is about knowledge, it is about wisdom, it is about good and bad, it is about behaviour, it is about love, it is about truth seeking, etc.

Soul, the most complicated of the three, is the spark that triggers thoughts in a living being. We start forming body immediately when the mothers conceive. But we will start having the life only when the Soul, the spark that triggers the thoughts, enters the body. There are different opinions on the point when the foetus in the mother's womb starts having life which is the soul that triggers thoughts, in the already existing body. Some say it is between 40 to 48 days and some say, even later.

The interconnection between Soul, Mind and Body is–Soul (spark) initiates (triggers) Mind (thoughts) in the Body (abode). Life is the body having mind characterised by Soul.

It is the origin of the Soul (spark) that characterises (determines the nature of) the mind (thoughts). The soul originates from the universe. It is nothing but the waves emitted by the celestial bodies. While the soul is absolute, the nature of the thoughts (mind) it triggers is determined by the position of the celestial bodies and the nature of the body that the thoughts are triggered in. The Soul is always around. The time of the entry of the soul into the body is determined by the readiness of the body.

However, the nature of the thoughts triggered by the soul is determined by the body they are triggered in and the position of the celestial bodies from which the soul emanates. That is the very reason that mother and the son are likely to think alike, siblings are likely to think alike, people born in the same month are likely to think alike. The nature of the thoughts continues to change as we grow, as we receive inputs from the education, from the incidents around and from our observations.

Let us dwell more on thoughts here. We all know thoughts occur and reside in our brain. However, nobody so far could explain satisfactorily how thoughts occur in our brain. Let us take a moment and one more attempt to understand how thoughts occur.

Thoughts are nothing but the electric triggers in a set of neurons in our brain called the process-neurons. To understand this further, let us understand the neurons, nervous system and brain.

Nervous system is responsible for sensing, processing and actions in and from our body. Nervous system is nothing but the network of neurons. Spinal cord and Brain form most important parts of the nervous system. Brain contains more than 28% of neurons of the entire nervous system. The number of neurons that brain contains is estimated to be more than 90 billion. Neurons are specialized cells of the nervous system that transmit signals throughout the body. Neuron is only in two states that is, active when the electric potential of the neuron is 30mv and inactive, when the electric potential of the neuron falls to -55mv.

When the neuron is active, it triggers other neuron connected to it to be active. Thus the signal passes from one neuron to the other. The signal in a neuron is always the same irrespective of the stimuli which are light and any type of its source, sound and any type of sound, smell and any type of smell, taste and any type of taste, touch and any type of touch like cold, hot, pain, soothing, etc.

What makes the difference between two stimuli like, two types of smells, two types of sounds, two types of objects that we see; is not the type of signal that the stimulus produces (they do not produce different signals as explained above) but the set of neurons (the neural network) that the stimulus activates.? When we see an object, say a person for the first time the retina in our eye activates a set of neurons (neural network) that spreads to brain as sensory neurons, in the brain as process neurons and back to muscular system as motor neurons. When we see the same person next time, the same set of neurons (neural network) is activated and thus, we recollect the person and say "Hi" to him. This is how our memory also works. Same way we sense a smell, sound or touch and remember them; and react to them as well.

The neural network that a stimulus triggers contains three parts, viz. the sensory neural network, the process neural network and the motor neural network. Sensory neural network triggers a set of process neurons (process neural network) which trigger a set of motor neurons (motor neural network) which cause movement in the muscles of our body. This is how we react to stimulus like saying "Hi" to the known person on seeing him, run away from danger or dance to rock music.

Our body muscles are moved by the motor neurons as triggered by our process neurons on the activation by the sensory neurons. The combination of the sensory neurons that trigger the combination of process neurons and that combination of process neurons that trigger the combination of the motor neurons initiate a specific movement in the muscles. It is this combination of the process neurons which is triggered, called thought. We can change the combination of process neurons (thoughts) to obtain desired movement in the muscles which is nothing but our action. Our thoughts are nothing but that network or combination of process neurons which stay triggered. When I have a particular thought, it means that a particular thought process neuron combination is active, which triggers particular motor neurons and hence, triggers a particular action from me.

Our thoughts decide our action irrespective of the stimulus. For example, a person absorbed in his own thoughts will not notice an object even though he sees it, will not hear a sound even though it is quite around, and will not even sense pain on hurt if he is absorbed in his own thoughts.

We all know and acknowledge that different people react differently for the same stimulus. The reason is the different thought process that they have at the time of stimulus that triggers different motor neurons which cause different reaction from them.

I had two colleagues in my company who joined the company together in the same grade. Both had same education qualification from similar standard of institutions. Both went through same selection process to get into the company. Both were handling almost the same profile of the jobs in the company. However, after 15 years in the company one was moving the ranks faster, the other one was lagging behind. The one who was moving fast was an aggressive and a different thinker and so were his actions, reactions and the one who was lagging was cautious and a conventional thinker and so were his actions and reactions. It was the thought process that differentiated the two.

The process of being aware of one's thoughts at the particular moment in time, understanding which of them are appropriate and effective, which of them are inappropriate and ineffective, controlling the inappropriate and ineffective thoughts and generating more and more appropriate and effective thoughts sets the differentiator in person to person and in performer to failure. This is what is called–Thought Intelligence.

With the help of Thought Intelligence, one can control the thoughts and thus the actions. Very high quotient of Thought Intelligence leads to exemplary actions. Leader shall possess high quotient of Thought Intelligence.

Some of the Positive thoughts, Negative thoughts, Poor thoughts, Right thoughts and Wrong thoughts are listed below for the ready reference of the reader. However, the reader can prepare his own list of positive thoughts to watch about.

Positive thoughts:

These include–Shall do good, shall help, shall love, shall develop, shall contribute, shall grow, shall work, shall follow guidelines, shall follow law, shall respect fellow beings, shall be grateful, shall express gratitude, shall obey superiors and elders, shall trust others, shall not be selfish, shall not hurt others, shall not belittle others, and so on.

Negative thoughts:

Harm, hatred, Being Jealous, never minding harming, never minding stealing, never minding cheating, never minding lying, never minding flouting, never minding misuse, never minding not being sincere, never minding selfishness, never minding hurting the feelings of others, never mind misusing, etc. include negative thoughts.

Poor thoughts:

Poor thoughts are the third type of thoughts which are neither positive nor fall in the category of the negative thoughts, but are poor in nature like – having no need to help, no need to follow guidelines, law and regulations, no need to contribute, no need to work, no need to love, no need to be grateful, no need to express gratitude, no need to obey superiors, feeling nobody is superior to me, I am only great, Whatever I do is right, I cannot go wrong, being over-proud, feeling others are not trust worthy, not being sincere, having no care for the feelings of others, preaching but not following, and so forth.

Right thoughts and Wrong thoughts:

Wrong thoughts are either negative thoughts or even if they are not negative, they are not apt thoughts for the occasion. Deciding which is right thought and which is not depends on the occasion of the task. It requires job skills of the task to know the right or wrong of the thought. One who has the right thoughts succeeds in the task.

It is the essential part of thought intelligence that one shall be aware of; the thoughts one is having, knowing which of them are right and which of them are wrong, knowing how to develop the right thoughts and how to deal with the wrong thoughts and leading oneself to a state of having all right thoughts (right is both positive and useful, wrong is either negative or not useful or both). The following concepts will help in harnessing positive thoughts in oneself-

Thoughts are Attitude

It is appropriately said that attitude is the thought process with which one approaches a task. Results comply with approach. Result is the direct possibility of the approach and approach is the function having the variable as thought process. If we approach with the thoughts of success, the possibility is success. If we approach with the thoughts of failure, the possibility is failure.

Result is the function of approach with thoughts as variable.

$$R \propto A(T_p)$$

$$\text{Result} \propto \text{Approach (Thought process)}$$

Where T_p is thought process

If thought process T_p is success result is success.

If thought process T_p is failure result is failure.

The thoughts that determine the attitude of a person are called attitudinal thoughts which are succeed-fail, smooth sail-road blocks, confidence-doubts, trust-mistrust, capable-not capable, accommodate-burden, useful -waste, right-wrong, friend-foe, contributor-competitor, always-never, accommodative-selfish, humble-arrogance, enthusiastic-lack lustre, etc.

Leader always approaches the task with the right attitude and gets the desired results. Leader demonstrates the right attitude always and makes it essential part of his habits.

Positive look

A manager was asked by the top management to study the company, its shortfalls and prepare a report suggesting the improvements. The manager prepared the report like this-

Our company is great. But it lacks the spirit to face the competition. Our employees are de-motivated a lot. Our product lacks quality. Our plants are very old. The HR department does not accord proper trainings. Finance department does not approve notes for buying new equipment. We do not have good department heads. The current HODs put lot of pressure on the subordinates. Our marketing model is outdated. Our channel members are lethargic. The customers still have lot of faith in our services. The customers show lot of loyalty to our brand. We may lose this soon as we are not putting any efforts to retain the customer faith.

We suggest a complete revamp and shaking. It is suggested our plants be demolished and new plants be commissioned. Our HR department shall be punished if it is not providing sufficient training. The old people in our finance department may be removed and put in some less significant department. Our product quality and design department must learn from our competitors. There shall be transparent complaint redress system to encourage subordinates who are suffering from the harassment of the bosses to open up.

The manager showed the report to his boss before putting up to the management. The boss studied the report and suggested his manager to rephrase the report by avoiding using the negative words without distorting the facts. The quick to grasp manager prepared the report again and this time the report read like this-

Our company is great. The spirit it had in facing the competition lies endangered lately. Our employee motivation requires renewed vigour. The HR department is being helpless in according proper trainings. The lack of quality in our products requires urgent attention. Our plants have grown old. Finance department has constraints in approving notes for buying new equipment. We shall have good department heads who distance themselves from archaic practices like subjecting the team members to lot of pressure to get the work done. We shall have the marketing model suiting to the current market conditions. The lethargy in our channel members shall be driven away. The customers still have lot of faith in our services. Impressive steps on war foot shall be taken to retain the customer faith from losing.

We suggest a complete revamp. It is suggested our plants which were our strength in the past as per the then requirements be reconstructed as per the current requirements. Our HR department shall be incentivised to provide sufficient trainings. The wisdom of the senior personnel in our finance department shall be taken advantage of in the roles like consultation and patron. Our product quality and design department shall be given regular opportunities to learn from our competitors. There shall be transparent feedback system to encourage our junior colleagues to provide open and frank feedback about seniors and peers.

The revised report can win wider acceptance as it is against no body. Most of the negativism is left out without distorting the facts. It requires a careful and meticulous practice to avoid negativism. It is very important to ensure that our thought process is not wrought with negativism and our sentences do not include negative words.

Thought Intelligence is an important leadership virtue. Thought intelligence provides purity of thoughts, control and alignment of thoughts, positive thinking and outlook. A leader is required to have very high thought intelligence quotient. Thought intelligence is the basis for leadership value system and other leadership values, virtues and actions stem from thought intelligence.

Thought intelligence shall be inculcated in a person from the childhood without linking to leadership. Parents and teachers shall encourage and provide the guidance to children to develop Thought Intelligence at the early state. One basic process to improve the thought intelligence quotient for leadership engineering is explained below-

Emotional Intelligence

Emotional Intelligence is another important part of the 3CVA qualities that forms the leadership habit system. Great works by eminent personalities like Daniel Goleman have opened up the importance of emotional intelligence in a person's success. Emotional intelligence is being aware of one's emotions at the particular moment in time, understanding which of them are positive and which of them are negative, controlling the negative emotions and generating more and more positive emotions.

Let us dwell more on understating the emotions. There are basically three types of motions in our body-

- Keymotions
- Emotions
- Imotions

Keymotions in the body are keys to life like digestion, breathing, heartbeat etc. The Keymotions are involuntary. These keymotions are caused by motor neurons alone without any stimulation, like heart does not wait for any stimulation to beat. There is no involvement of sensory neurons and thought neurons in keymotions.

(There are basically three types of neurons in human nervous system. 1) The sensory neurons which carry the sensory information to the central nervous system, 2) The motor neurons which cause muscle movement in our body, 3) The thought (process) neurons which process the sensory information and activate motor neurons accordingly.)

Emotions are the basic and first bodily reactions to the sensory inputs like fear, angry, happiness etc. The emotions are semi-voluntary. Emotions are caused by motor neurons on activation by sensory neurons. There is no involvement of thought neurons.

Imotions are intelligent motions. Imotions are voluntary.

Imotions are like thinking and then acting. Imotions are caused by motor neurons activated by thought neurons on processing the information from the sensory neurons.

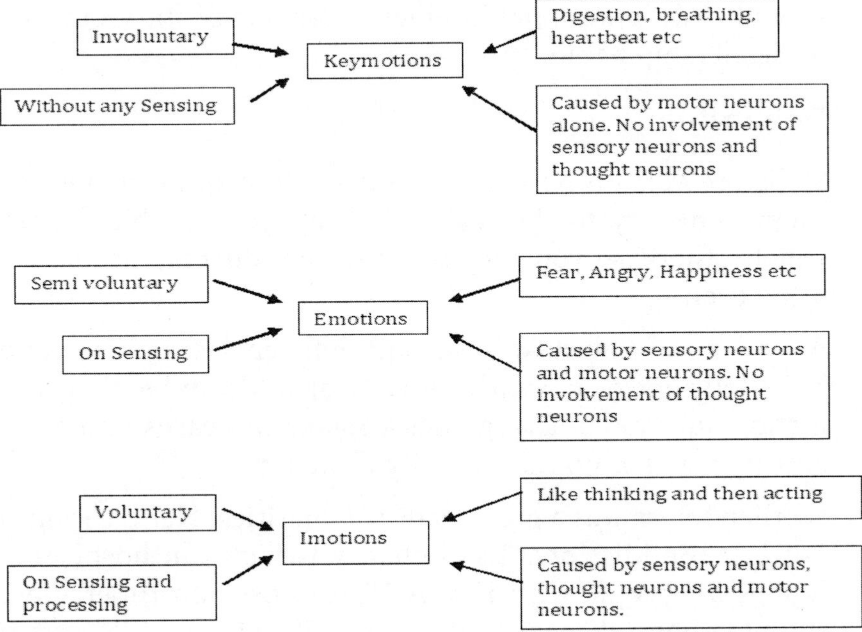

An emotionally intelligent person will have more of Imotions. Imotions become semi-voluntary motions, replacing the emotions for a leader and turn out to be habits for the leader.

Emotions like thoughts push a person more and more in the set direction, be it positive or negative. Emotions unlike thoughts in a person are evident and contagious. They fill the environment around. Positive emotions like joy, smile, pleasure, calm, cool composure, etc. push a person to be more and more like and fill the surroundings with the same aura. Negative emotions like sadness, anger, grudge, tension, jealousness, crying, shouting, hurrying, temperamental, etc. also push the person to be more and more like and fill the environment around with the same temper.

An emotionally intelligent person is the most liked and has very high success rate. His social acceptance is very high. The leader should make emotional intelligence as one of the elements of his leadership habits.

DoSo Skills

Skills are very important. One need to acquire if one lacks them. Sincerity to the task will make you exhibit the skills which otherwise you may not have or you may not show in other events.

A leader can afford to have skill deficiencies. But he cannot have deficiencies in values, virtues, attitude and actions for, it is the value system that guides a leader and earns him his due and even makes up for the skill deficiency.

Mother Teresa was a great leader of charities. Her creation was not just the Missionaries of charity (which run hospices and homes for people with HIV/AIDS, leprosy and tuberculosis), soup kitchens, dispensaries and mobile clinics, children's and family counselling programmes, orphanages, and schools; her creation was a sense of pride and purpose in the millions of destitute. Her creation was a mission which was to care for, in her own words, 'the hungry, the naked, the homeless, the crippled, the blind, the lepers, all those people who feel unwanted, unloved, uncared for throughout society, people that have become a burden to the society and are shunned by everyone'. Her creation was a beautiful death for the people who lived like animals; to die like angels—loved and wanted.

Her creation was a renewed surge of service to the people and to show them the ways and means. She did not base her creation on any skills. She did not have to. It is a huge creation no mere skilful person can ever create. Mother Teresa's this creation was based on the values, virtues, actions, attitude and leadership vehicles–on the values and virtues of sacrifice, charity, help, love, service; acts of courage, inspiration, sense of fulfilment; attitude of hope and purpose; leadership vehicles of wish, interest, aspiration, clarity, passion, urge when she left her family at the age of eighteen and never returned to her mother and sister, when she extended her hand to the dying destitute, when she faced difficult times in her path.

Teresa wrote in her diary that her first year was fraught with difficulties. She had no income and had to resort to begging for food and supplies. Teresa experienced doubt, loneliness and the distraction of temptation to return to the comfort of convent life during these early months. She wrote in her diary:

"Our Lord wants me to be a free nun covered with the poverty of the cross. Today, I learned a good lesson. The poverty of the poor must be so hard for them. While looking for a home I walked and walked till my arms and legs ached. I thought how much they must ache in body and soul, looking for a home, food and health. Then, the comfort of Loreto [her former congregation] came to tempt me. 'You have only to say the word and all that will be yours again,' the Tempter kept on saying ... Of free choice, my God, and out of love for you, I desire to remain and do whatever be your Holy will in my regard. I did not let a single tear come".

Leadership can be without skills, but not without leadership vehicles, values, virtues, actions and attitude. Skills will help only after having the value system. We know many leaders like Walt Disney, Soichiro Honda, Thomas Alwa Edison, Abraham Lincoln, and numerous other leaders who could not impress with their skills initially and had failure to begin with. They went ahead with leadership vehicles and mastered the skills required to succeed.

A leader in the process of leadership looks at two types of skills-

1) Domain skills

2) Soft skills

Domain skills:

Domain ability is complete understanding of the field in which a leader is working and mastery on important aspects of the field. Leader shall have clear opinions and knowledge on happenings. Leader shall learn constantly and upgrade himself with domain knowledge and abilities.

Soft skills:

Communication, Speaking skills, Listening skills, Emotional Intelligence, Presentation skills, articulation, Networking, Interpersonal skills, (Body language interpretation and reaction practice,), Learning from observation, Time management, smart working, motivation self and others, analytical ability, negotiation, fast reading, giving importance for specifics, speed, concentration, thinking on the feet, smile, walk, and looking into eyes form part of soft skills.

One shall not bother himself too much about the correct categorisation of the elements of leadership habits like if empowering is virtue or a skill or a thought. Whatever is the categorisation of these elements they form part of leadership habits.

Some tips are provided below that help inculcate leadership habits.

Tools to Acquire Leadership Habits

Consciousness process for improving Thought Intelligence:

Developing thought intelligence is easier said than done. Consciousness process will help improve the thought intelligence quotient. Consciousness process is practicing being conscious and knowing right and wrong. Consciousness process can be done off-the-task and on-the-task. Following are the steps in consciousness process.

Off- the- task consciousness process:

- Spend 16 minutes every day as per the 8hr16mnts norm to review the thoughts you had in the day.
- Jot down the thoughts and tick which are right and which are wrong.
- If you feel you are not being fair in judging the right and wrong, use the table below.
- Decide in the mind that you will not have these negative thoughts next day.
- Even if one cannot resist a thought initially, telling oneself that the thought was a negative one helps continue the process.
- Do the off-the-task consciousness process for 30 days to start the on-the-task consciousness process.
- On-the-task consciousness process:
- Being conscious of the thought as it comes is on-the-task consciousness process.
- Analyse the thought real time to know whether it is right or wrong.
- Decide on the very moment to not to have the thought next time.

One sees only what one looks at

This world has everything around us; useful–waste, good-bad and so on. But one cannot see all. Each one of the things available in the world also has everything in that. One sees it as one looks at it to be. Some people call one person good but another set of people call the same person bad. The later looked at bad in the person and they saw only bad. The first looked at good and the saw only good in him.

Look around for good or what is useful; you will start seeing the same. Leader looks around for opportunities and he sees opportunities. If you look from the angle of issues, you will end up seeing only issues.

You always have time for what you want to do (Time Management)

"I am not having sufficient time to do the task"–This statement is always heard by us, said by most of the people. Obviously, you cannot have time to do everything. Time like any resource is limited.

But you always have sufficient time to do what you want to do. The same people who say they do not have time to do a particular task say so, after allocating time first to do what they want to do. So the catch is you will not find time only if you whole-heartedly do not want to do a task. You always get time for your priorities. The leader is very clear on his priorities, so he always has time to attend to them.

You are being watched:

One shall not think he/she will display leadership values, virtues, attitude and actions only when occasion rises. It is wrong to think that these are meant to display only in the case of some events, meetings or occasions.

One is being watched all the time and impressions are being made about him all the time, whether off the task or on the task, whether in the formal events or informal gatherings. You are being watched and evaluated all the time. Opportunities open up, teams are built and success is determined sometimes accordingly, without one's knowledge and without any specific effort.

Leadership by default displays the values, virtues, attitude and actions all the time. They become part of their habits and hence, these are called leadership habits.

Meditation

Meditation helps thought intelligence. Meditation is negotiating with the thoughts. Negotiating with the thoughts is being aware of all the thoughts, deciding on which are useful thoughts, focussing on useful thoughts by keep shunning the other thoughts. Meditation is deep thinking on a subject with a stable mind.

Meditation is a continuous process. It can be done while walking, eating so on and so forth for the brain never stops having thoughts.

Check for Walk the Talk:

Actions shall be consistent with words. Talk the walk is easier than walk the talk. The basic reason some people command lot of respect and some do not is that the actions of the latter do not follow their talk. They do not walk their talk.

This was my boss in one of my earlier companies. For a call from a client, he promised a response the very next day. I was also part of the discussion. There was no response the next day from my boss to the client. The client reminded. My boss made commitment of response again for the next day. He did not keep his commitment. The client never returned. Eventually, we learnt it was a huge potential client and we lost an opportunity. This incident in my early career taught me two important lessons. It is very important to follow your words with actions. Failing which the trust and the confidence of the other person in us is completely lost. Our trustworthiness comes at stake.

It is also like we ourselves not respecting our own words. The second lesson is, do not commit what you cannot do if at all you cannot do. Saying "yes" is very easy. Saying "no" is very difficult. Saying yes and not doing is catastrophic. Nobody would like to hear a "no". But what is more damaging is saying "yes" and not doing. It is equally important to learn to say "no" to what you are not capable of as to gain the capabilities of following up "yes" with actions.

Check for Consistency:

What is more important than doing hundred things one time is doing five or six things hundred times. Consistency is having the repeatability. Consistency improves predictability. Consistency itself does lot of communication. It becomes difficult for one to understand what the next action of the leaderee will be if actions lack consistency. Leaderee shall bring in consistency in all his walks and talks.

Check for Excellence

Excellence is the power with simplicity. The best example for the statement is the current-day smart phone. The smart phone is very simple to use and yet, very powerful tool fitting just in the palm. It took many years for the man to achieve this excellence in communication technology. Leaderee shall thrive at all times for simplicity and clarity in his actions and accomplishments to achieve excellence. After all, excellence is power with simplicity.

LEADER SOCIETIES

Why some countries and some provinces in some countries prosper, bustle with activity and experience remarkable growth and others do not? Some institutions and companies achieve remarkable things and others do not. Look at the nations, provinces and places like USA, Japan, Singapore, Taiwan, UK, Germany, Silicon Valley, Hongkong, etc. These nations and places have transformed themselves and transformed the mankind. How is it possible for USA to produce leaders from Franklin to Zuckerberg, How it is possible for Japan to transform itself and transform industries like Electronics, Automobile, Robotics? How it is possible for Singapore, a tiny nation to transform itself into a major economic force in the world?

Look at companies, institutions, organisations like Toyota, Stanford, ISRO and so on. Their contributions to their industries and thus, to the world and for the progress of the mankind is enormous. These nations/ provinces/places and companies/ Institutions/organisations are truly leader societies (nations/ provinces/places and companies/Institutions/organisations are commonly referred to as societies hereafter for simplicity in understanding).

What makes these as leader societies is the abundance of leadership ambience and leadership spirit in them. This leadership ambience and the leadership spirit enable continuous creations and continuous emergence of leaders.

The leadership ambience which is highly prevalent in the leader societies is the ambience of having no leadership barriers. It is where ideas flow freely, it is where there are no or minimum early leadership life struggles, It is where information is available freely who ever seeks, It is where the constraint is only the thought of the task. It is where there is no social stigma attached to failure and there is no fear of failure. It is where the person knows his interests, surpasses his ignorance of the forward path very early. It is where there is no social insecurity and it is where the preferences are not forced upon him.

Leadership spirit is the spirit of liberty and prosperity. Any leadership act as the history tells, has sought fundamentally two things which are, liberty and prosperity and it thus ended up in transformation. The spirit of liberty is fundamentally the liberty from the four walls of ignorance, liberty from the narrow mindedness, liberty from the oppressive mindset and superstitious thinking. The spirit of prosperity is making today easier and better than yesterday.

In the leader societies, the direction is always the leadership spirit. By nature, it becomes the basic character of the members of the leader societies to live happily and contribute to the liberty and prosperity of the society that is, to him and to his fellow member citizens.

On the examination of the leader societies, you will find the leadership ambience and leadership spirit as described above in abundance. Any society or institution shall thrive to have the leadership ambience and the leadership spirit in them. Any society that lacks or shuns the leadership ambience and leadership spirit is retrogressive and mediocre.

ISRO, the leader organisation:

ISRO was founded by the great leader and the father of Indian Space Programme Mr Vikram Sarabhai. Prof. Satish Dhawan took the mission at ISRO forward. With the spirit and ambience set up by Prof. Dhawan, ISRO kept creating in the Indian Space programme and research and transformed India in the sectors like communication, and that of climatic. ISRO kept building its satellites, launch vehicles, capabilities and technologies and stood up to be the pride of the nation and set examples for the third world. It created history with its Moon and Mars missions. It created history by launching 104 satellites aboard a single launch vehicle. The heads of ISRO kept the spirit sustained in the organisation enabling ISRO to be the leader organisation.

The general process of building the leader societies contains a strong leader envisaging and creating such societies along with and as part of them, the leadership ambience and the leadership spirit. Like in the case as above, like in the case of Toyota by Shakihiro Toyoda, like in the case of USA and the founding fathers, and so on. By the very nature of this ambience and the spirit, the heads of the leader societies that follow, will keep up the leadership ambience and leadership spirit and ensure the sustenance of leadership. There will emerge other leaders time to time to fill leadership gaps in such societies.

Each society shall strive to be a leader society. Leader society does not, at all, mean being the first in the race. Leader society is a society that creates continuously. Each society shall strive to have the leadership ambience and Leadership spirit integrated in it.

It is the foremost responsibility of the society, to become a leader society, to remove the barrier of 'Ignorance and Early Life Struggles' in the Leadership Life Line. 'Ignorance and Early Life Struggles' is the first and the toughest barrier in the Leadership Life Line. These barriers have the crushing effect on the leadership. These barriers won't let the leaderee even start. The reader may refer back to the earlier section to recollect the nature of this barrier.

Leader Societies shall create such surroundings around that keep triggering 'interest' in their members. It may be recalled that interest is the leadership vehicle that allows leaderee to overcome barrier of 'Ignorance and Early Life Struggles'. Let us explore how this can be done in a society in the context of nation/province/place.

Hubs:

The society, that is nations, provinces and places in this context, shall create sector-specific hubs at different places for different basic nation building sectors like Manufacturing, Trading, Technology, Healthcare, Culture and Fine Arts, Education, Spiritual, Political Administration, Social Service, Sports, Combat Services, Tourism, Journalism, Financial Services.

The society shall concentrate all the activities pertaining to the sector at these hubs. Society shall create basic infrastructure for the development of these sectors at these hubs. Individuals and the institutions (companies, organisations, agencies) interested in this specific sector will gather at the hub. The ambience and the surroundings of the hub get filled with activities related to the sector. The basic infrastructure coupled with the concentration of the 'Interest' will remove the 'Early Life Struggles' in the leadership process.

With the system of hubs, the barriers like differing preferences, ambiguity, and hardships get addressed. Information is available readily at a hand reach. Clarity improves. All like-minded people gather together. Differing preferences are seen very less.

Elimination of Fear of Failure:

Fear of failure stifles creativity. It will not let tread the path unknown. Fear of failure forces one to limit himself to comfort zone. Societies with the structural and cultural changes shall ensure subsidization of the stigma attached to failure.

Societies shall ensure sufficient social security to encourage people to take risks. Social security to ensure a failed person is not crushed economically, socially but rises smoothly from failure. The recovery happens when there are enormous opportunities. A person failed shall not be seen as untouchable. A failed person shall be seen to be equally good like any other person. In fact, failed person is more valuable as he has lot of experience and will work with all the caution. However, a habitual failure shall not be equated to sincere failure. Habitual failure is one who does not learn lessons from the failure; there is no Failure Incident Analysis and thus, fails constantly and several times. The sincere failure learns all his lessons and puts them to implementation in his next venture.

Consistency in the vision and policies for better communication:

Communication of the idea of building a Leader Society shall go deep into masses. It is important, for this purpose, to maintain consistency in the vision and policies towards the purpose of building a leader society, irrespective of change in governments.

The talks, the discussions, the actions shall all be about the same vision and policies through out. For the communication to go deep and to be understood in its true sense, it shall talk the same thing repeated. Frequent changes in the subject will not let the communication reach deep and understood in its true sense. Frequent changes in policies will lead to confusion, chaos, change directions frequently and end up being directionless. The vision and policies shall be so strong that they will become the culture.

Recognition:

Recognition and appreciation is the natural craving of the mankind. Man seeks recognition and avoids humiliation. This is the basic principle which explains human behaviour. Recognition inspires. Societies shall recognise the leadership efforts by the individuals or institutions. It can be in the form of awards like that of micro leader, micro leader institute, leader, leader institute, leader hub, etc. No genuine effort must go unappreciated.

Leadership Engineering Centres:

Societies shall establish Leadership Engineering Centres. The Leadership Engineering Centres shall take up, in mass scale, assessment of the people for their interest and personality profile and encourage them to submit proposals in their area of interest. These centres shall provide support for the leaderees with ideas like validating their proposals, attaching them to investors and mentors and tracking the progress.

Leader Institutions:

Institutions are the pillars of a society. Societies to become leader societies will have to build numerous leader institutions. The leader institutions in the main nation-building activities like Manufacture, Trading, Technology, Healthcare, Culture and Fine Arts, Education, Spiritual, Political Administration, Social Service, Sports, Combat Services, Tourism, Journalism, Financial Services will constantly churn out leadership and contribute in the prosperity of the sector and thus, making the society a leader society.

The societies shall encourage people to establish institutions, provide support to build them as leader institutions by giving them the exposure, the responsibility and the clear definition of their role in the nation building. Institutions thus built shall aim to be leader institutions that create constantly. This is possible by engineering leaders and leadership and making leadership engineering the integral part of their culture and existence. The leadership engineering programmes in an institution can be as the following-

Leadership Engineering Programme

Leadership Engineering, as has been discussed in this book, is a very effective approach that helps meet the leadership requirements of the companies, societies and institutions with very high efficiency. Companies, societies and institutions shall have *Leadership Engineering* Programme (LEP) on the basis of concepts discussed in this book as integral part of their culture. Leadership Engineering Programme (LEP) shall be a three step process.

Step one is the Content Transfer in the form of workshops. The workshops shall have four modules. The workshops contain content presentation through participation of students with video/audios, case studies, group exercises, DIYs, experiments, discussions, etc.

Module one shall be about "Understanding leadership, characteristics of leadership and leadership requirements in a Corporate". At the end of the module, the participant shall get clarity on what exactly the leadership is and shall gain all the confidence to start his/her leadership journey. Module 2 shall be covering the topics on "Decoding Leadership Process" and "Identification of Leadership tasks". At the end of the module, the participant shall understand the different bottlenecks that he/she will come across in his/her leadership process. The participants shall be made to generate ideas and leadership tasks with the idea generation techniques as discussed in the book.

Module three will be on "Engineering a Leader". The participant shall be taken through the practical experience of leadership. At the end of the module, the participant shall be able to understand how to get over the bottlenecks in the leadership process with some practical experience. Module four shall cover "Leadership Habits" and "Leader Societies". The participants shall be made to identify the leadership habits and shall be provided with the techniques (as mentioned in the book) on how to make these habits their own habits.

Step Two is the selection of Leadership Tasks. Step two shall have the identified potential leaders (Leaderees) identify the leadership tasks (as defined in the book) with the framework provided in the book is "N-E-T-E-T-E" which is identifying the "**N**eeds/improvements"–examining the "**E**xisting setup"–identifying the "**T**ask" required to meet the need–proving the "**E**ffectiveness" of the task and – "**T**esting" the task with the help of various techniques and tools provided in the book. At the end of this step, the leaderee will be ready with the leadership task to execute the same.

Step Three is embarking on leadership process to complete the leadership task. Leadership process is the execution part of the leadership. The leaderee may fail or stop proceeding further at any leadership state in the leadership process due to leadership barriers which may appear insurmountable to him. The leaderee shall be provided with the constant guidance or shall be made to have self-guidance in identifying the leadership barriers and overcome the same with the help of leadership vehicles as enumerated in this book, to enable him move from one leadership state to the next and complete his creation.

Conclusion Is The Beginning Of The Next

Knowing leadership does not end with completing reading this book. It begins now. The very fact that you are reading these lines now (assuming that you have not started reading this book by reading the conclusion first) indicate that your resolve is unperturbed and you are ready for the next step. The next step is to select your leadership task and start your leadership process to see the leadership task reaches its conclusion. The world requires large pool of leaders and deserves much better leadership. You shall be one such leader.

This book will guide you at each and every step in the leadership process. Refer this book if you find yourself on a cross-road. Refer this book if you feel you have come to a jolting halt and do not know what to do further. Refer this book if you feel that you are being pulled away from what you desire to create. This book aims to be your constant companion.

The process of creation of this book from the inception of the idea till it's landing in your hands now today has followed the typical leadership process that it has enumerated in itself. It has seen all the possible barriers like denials, disbeliefs, cold responses (of course, lot of pats on the shoulder too) and has successfully crossed them all to see today, thus presenting, yet another example of leadership process.

This book is the creation which has just started creating–You.

Get going

****

INDEX

NOTES

NOTES

NOTES

NOTES

ABOUT THE AUTHOR

V Srinivasa Prasad is a marketing professional with more than eighteen years of experience in the marketing, retail and institutional sales and technical services in petroleum sector (Lubricants). He is currently working with IndianOil Corporation Limited as Chief Manager.

The author's area of special interest is leadership. His opportunities of meeting people with different profiles that he came across in his extensive travels and his study of the contemporary and past leaders and his more than five years of research on leadership, helped enhance his understanding of leadership and leadership related issues, thus resulting in this book.

The author can be contacted at

author.leadershipengineering@gmail.com

The concept that leaders can be engineered, just like any other thing or process, is a revolutionary one. When countries like India are in dire need of leaders in every walk of life, this book is timely and relevant. Both organizations as well as individuals would immensely benefit from the simple and effective methods that the author has provided in this book to create leaders from common people.

Dr. Najeeb Kuzhiyil

Author of Spirit of Engineering;

Oil industry professional, Houston, Texas, USA